The Living Galaxy

Eugenio Mieli
Andrea Maria Francesco Valli
Claudio Maccone

The Living Galaxy

Winners and Losers in the Milky Way

Springer

Eugenio Mieli
Pomezia, Roma, Italy

Claudio Maccone
Istituto Nazionale di Astrofisica
Roma, Roma, Italy

Andrea Maria Francesco Valli
Société Scientifique du Bourbonnais pour
l'Étude et la Protection de la Nature
Moulins-sur-Allier, Allier, France

ISBN 978-3-031-67326-9 ISBN 978-3-031-67324-5 (eBook)
https://doi.org/10.1007/978-3-031-67324-5

Claude, a virtual assistant created by Anthropic, provided the language traslation and revision for this text. A subsequent human revision was done primarily in terms of content.

Translation from the Italian language edition: "La Galassia Vivente" by Eugenio Mieli et al., © The Editor(s) (if applicable) and The Author(s), under exclusive license to Springer Nature Switzerland AG 2024. Published by Springer Nature Switzerland. All Rights Reserved.

This Springer imprint is published by the registered company Springer Nature Switzerland AG
The registered company address is: Gewerbestrasse 11, 6330 Cham, Switzerland

Exploring the fields of astronomy, geology, biology, paleontology, and futurology, we find that the current number of planets inhabited by galactic civilizations like ours is estimated to be around 3, including Earth and perhaps one or two others located approximately 17,000 light-years away from us, a vast distance.

However, just as life forms are born in a specific niche and, if conditions permit, can spread across an entire ecosystem, a galactic civilization that overcomes all the challenges it faces could potentially spread throughout the entire galaxy, becoming, in essence, ETERNAL. In such a scenario, the present number of civilizations would exceed 2000, all highly advanced and capable of traversing the galaxy.

Thus, the question we should be asking ourselves is different: if we were a super-civilization with the ability to traverse the galaxy, where would we choose to venture? Would solar systems like our own hold any interest for these civilizations?

Foreword

Are we alone in the universe? This question has captivated the minds of scientists, philosophers, and dreamers for centuries. It strikes at the heart of our understanding of our place in the cosmos and the nature of life itself. As we seek to answer this fundamental question, the search for extraterrestrial intelligence (SETI) has become one of the most important and exciting fields of scientific inquiry.

Ultimately, the question can only be answered through observation. We must search the skies for signs of life beyond Earth, whether in the form of radio signals, biosignatures in the atmospheres of distant exoplanets, or even direct evidence of the technology employed by extraterrestrial civilizations. However, while we continue to observe and explore, we can also use the Drake equation to guide our thinking about the likelihood of life existing elsewhere in the universe.

The Drake equation, first proposed by astronomer Frank Drake in 1961, is a probabilistic argument for estimating the number of active, communicative extraterrestrial civilizations in our Milky Way galaxy. The classic form of the equation takes into account various factors, such as the rate of star formation, the fraction of stars with planets, the number of habitable planets per star, and the likelihood of intelligent life emerging and surviving long enough to communicate. By assigning values to these variables, we can arrive at an estimate of the number of civilizations that might exist in our galaxy.

Some have called the Drake equation a way of quantifying our ignorance about the prevalence of life in the universe. After all, many of the variables in the equation are currently unknown or difficult to estimate with any certainty. However, the equation serves as a valuable framework for thinking about the factors that might influence the emergence of life and intelligence beyond Earth.

In their groundbreaking book *The Living Galaxy*, the authors take an original and exhaustive approach to the Drake equation. They delve deep into each of the equation's variables, drawing on the latest research from a wide range of scientific disciplines. From astronomy to anthropology, geology to genetics, the authors explore the cutting-edge science that is shedding new light on the possibility of extraterrestrial life.

Readers of *The Living Galaxy* will be exposed to a wealth of fascinating insights and discoveries. They will learn about the ongoing search for exoplanets, the potential for life to emerge in the harsh environments of other worlds, and the evolutionary processes that might lead to the development of intelligence and technology. The book also explores the cultural and philosophical implications of the search for extraterrestrial life, and how the discovery of alien civilizations might change our understanding of ourselves and our place in the universe.

As you embark on this journey through the pages of *The Living Galaxy*, I invite you to keep an open mind and a sense of wonder. The search for life beyond Earth is one of the great scientific adventures of our time, and the Drake equation remains a useful tool for guiding our thinking and inspiring our imagination. Whether we are alone in the universe or part of a vast cosmic community, the quest for answers is sure to be a thrilling and enlightening one.

Lee-on-the-Solent, UK Stephen Webb

Introduction

This is not merely a *scientific* book. Rather, it aims to primarily tell a story of a series of events, beginning with the pre-cosmic era, progressing through prehistory, and culminating in a futuristic narrative populated by vivid scenarios and powerful protagonists. Although their existence is rooted in scientific models, together they represent the drama of life itself – being born, imposing its presence, or dying and disappearing.

The setting would be akin to many mythical theogonies, the theme equally transcendent: the creation of the world, the fate of the beings that inhabit it, the fall and potential redemption, the initial reality, and the final one. However, the singers of such an epic poem are now astrophysics, geology, chemistry, biology, paleontology, and futurology, and the poetic meter they employ is mathematics.

The action commences among the last stars, sisters indeed, but profoundly different from one another. It continues in planetary systems that, much like diverse ecosystems, prove either hospitable or hostile to life. It witnesses life's inevitable emergence as soon as conditions permit. It follows life's various attempts to preserve itself, evolve, and push towards intelligence and technology. It bears witness to an intelligence exposed to the greatest risks – those created by itself – which threaten to plunge it back into the darkness of extinction. It concludes its journey alongside a handful of survivors, superior beings, true citizens of the galaxy free to roam among the stars.

A story of possibilities, attempts, defeats, and redemptions, on the scale not of humankind, but of galactic civilizations.

Contents

1

Where Is Everyone?

In 1950, during a lunchtime conversation discussing extraterrestrials, the renowned physicist Enrico Fermi turned to his colleagues Edward Teller, Emil Konopinski, and Herbert York, and asked a thought-provoking question: "Where is everybody?" His colleagues were accustomed to Fermi's penchant for mentally calculating solutions to highly complex situations from what seemed like insufficient data. In this case, he had mentally estimated the number **N** of contemporary civilizations that should populate our galaxy, arriving at a certainly high value, despite the lack of evidence of their existence—a phenomenon known as the "great silence."

This paradox, now referred to as *the Fermi paradox*, remained unanswered for a decade.

In 1960, the first **S**earch for **E**xtra**T**errestrial **I**ntelligence (**SETI**) was conducted by Frank Drake, who shortly after listed the numbers we would need to know to estimate the number N of communicating societies in the galaxy [21]. In reality, this list was originally intended to serve as an agenda for a meeting at the Green Bank Observatory in West Virginia. However, his list was perceived as a genuine equation, a product of seven factors that could be used to calculate **N**. These factors were:

$$\mathbf{N} = \mathbf{R}^{*} \cdot \mathbf{f}_{p} \cdot \mathbf{n}_{e} \cdot \mathbf{f}_{l} \cdot \mathbf{f}_{i} \cdot \mathbf{f}_{c} \cdot \mathbf{L}$$

$\mathbf{R}^*$ is the annual rate of star formation in the Milky Way
$\mathbf{f}_p$ is the fraction of stars with planets
$\mathbf{n}_e$ is the number of planets suitable for life per star

E. Mieli et al., *The Living Galaxy*, https://doi.org/10.1007/978-3-031-67324-5_1

f_l is the fraction of suitable planets where life develops
f_i is the fraction of planets inhabited by intelligent life
f_c is the fraction of planets where intelligent life decides to communicate
L is the lifespan of the planet in which intelligent life persists

Drake had probably written down what Fermi had simply thought a decade earlier. However, it was soon realized that this first description was insufficient to reach concrete conclusions since the seven factors often fluctuate between very distant minimum and maximum values, and in some cases are completely unknown. We refer especially to the factors from the fourth onward, where astronomical knowledge is no longer of much help [114].

In 2008, Claudio Maccone, while chairing the Permanent SETI Committee of the International Academy of Astronautics based in Paris, devised a powerful mathematical tool called *the statistical Drake equation*. This equation, starting from the statistical distribution of the seven individual factors, obtains the statistical distribution of their product according to a curve called *lognormal*, or the distribution curve of a random variable whose logarithm is distributed as a normal distribution [65] (Appendix **D**).

This result is not taken for granted and becomes increasingly accurate *as more factors contribute to the calculation*, as the error is reduced as the number of elements in the product increases (central limit theorem). A graphical example of the lognormal function **Φ** is shown in Fig. 1.1 while its analytical form is presented in Fig. 1.2.

As done by other authors, we have preferred to refer to a variant of the classic Drake equation that is more suitable for our purposes.

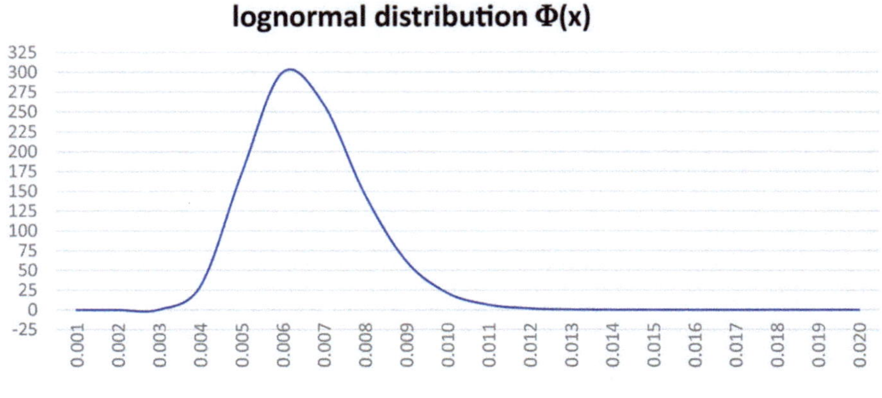

Fig. 1.1 Example of lognormal distribution **Φ**: the distribution exists for **x > 0** and is normalized with an average value, in the example given, around **0.007** and a deviation around **0.002 (IJA 14/06/2023 Mieli, Valli, Maccone)**

$$X_0 = \prod_j X_j$$

$$\Phi(X_0) = \frac{1}{X_0} \cdot \frac{1}{\sqrt{2\pi}\sigma} e^{-\frac{(\ln(X_0)-\mu)^2}{2\sigma^2}}$$

$$< X_0 >= e^{\mu} e^{\frac{\sigma^2}{2}}$$

$$\sigma_{(X_0)} =< X_0 > \left(e^{\sigma^2} - 1\right)^{\frac{1}{2}}$$

$$\mu = \sum_j \frac{B_j\left(lnB_j - 1\right) - A_j\left(lnA_j - 1\right)}{B_j - A_j}$$

$$\sigma^2 = \sum_J (1 - \frac{A_j B_j (lnB_j - lnA_j)^2}{(B_j - A_j)^2})$$

Fig. 1.2 General case of application of the lognormal **Φ** to the group of Uniform random variables $\mathbf{X_j}$ between the minimum and maximum values $\mathbf{A_j}$ and $\mathbf{B_j}$. This results in the distribution of the product variable $\mathbf{X_0}$ with mean $\mathbf{<X_0>}$ and variance $\mathbf{\sigma(X_0)}$ **(IJA 14/06/2023 Mieli, Valli, Maccone)**

Or, in our case:

$$\mathbf{N = N_s \cdot n_p \cdot f_s \cdot f_l \cdot f_i \cdot f_c \cdot f_L}$$

With:

- $\mathbf{N_s}$ number of galaxy stars suitable for life (i.e., of spectral class **K**, **G**, and **F**)
- $\mathbf{n_p}$ number of planets per star in the habitable zone (of spectral class **K**, **G**, and **F**)
- $\mathbf{f_s}$ fraction of stable planets in the habitable zone (function of duration $\mathbf{\Delta T}$)
- $\mathbf{f_l}$ fraction of suitable planets where life actually develops
- $\mathbf{f_i}$ fraction of planets inhabited by intelligent life
- $\mathbf{f_c}$ fraction of planets where intelligent life decides to communicate
- $\mathbf{f_L}$ fraction of the planet's lifetime in which intelligent life persists compared to the duration of the last stellar population (**about** 7 **Gy**)

As you can see, in the original Drake equation, the first factor is the annual star formation rate in the Milky Way, **R***, which is equal to $\mathbf{N_s/\Delta T_0}$, where $\mathbf{N_s}$ is the number of stars and $\mathbf{\Delta T_0}$ is the average duration of the last stellar population (7 **Gy** or 7 **billion years**). However, in our equation, the first factor loses the denominator $\mathbf{\Delta T_0}$. Instead, $\mathbf{\Delta T_0}$ divides the last term **L**, which becomes $\mathbf{f_L}$ (the fraction of a planet's lifetime in which intelligent life persists compared to the duration of the last stellar population). In this way, as we will see, we will be able to use the first terms of the Drake equation to directly obtain information not only about advanced civilizations but also about life forms in general.

Moreover, the choice was made not to delegate all the calculation effort on the third of the first three parameters of Drake's original equation ($\mathbf{n_e}$, or the number of planets suitable for life). Instead, we partially redistributed the work across the first two parameters: the first parameter, $\mathbf{N_s}$, has become the number of stars in the galaxy that belong *Only* to the spectral classes K, G, and F, which are suitable for the development of life (a star's spectral class is assigned based on its surface temperature [61]).

Similarly, the second parameter has become $\mathbf{n_p}$, or the number of planets per star in the habitable zone, thus also taking on the burden of discriminating *Only* the planets in the so-called "Goldilocks zone," where conditions are appropriate (remember that the parameter $\mathbf{f_p}$ in the original equation simply represented the fraction of stars with planets).

As a result, in our equation, the third parameter, $\mathbf{f_s}$, replacing $\mathbf{n_e}$ (the number of planets suitable for life), specifically becomes the fraction of stable planets in the habitable zone, which is also a function of the duration, $\mathbf{\Delta T}$, that we consider, depending on the stages of life development under consideration. As mentioned, this way of redefining the Drake equation will be useful both in estimating each individual parameter and in using the entire equation, which can be employed not only for counting galactic civilizations but also for estimating the number of planets on which life develops at different evolutionary levels.

Finally, while Drake referred the onset of intelligence to the fifth parameter and made it technological and communicating in the sixth, we have instead chosen to define intelligence according to a stricter criterion of energy availability, as we will see in detail, based on the model of Kardashev [51]: intelligence is technologically at least on our level, with a minimum parameter **K = 0.7**. This factor defines the perimeter of the fifth parameter, while the sixth takes charge only of civilizations that *Decide* not to communicate and remain in the shadows. The seventh, dealing with the duration of civilizations, also absorbs within it the cases in which the aforementioned civilizations *Are Induced or Forced* not to communicate anymore.

If we limited ourselves to a simple revisitation of the Drake equation, even with a new statistical tool like Maccone's lognormal, we would make very little further progress towards the knowledge of extraterrestrial civilizations. But we also established a crucial fact: the statistical result of a product is all the more accurate *the more factors contribute to the calculation*. This aspect of Maccone's algorithm allows us to fully immerse ourselves in the details of each single parameter without fear of losing precision but, on the contrary, with the certainty of gaining it. By doing this, in the rest of the book, we will move from estimating the probabilities related to the seven factors described in the equation to the following **50** factors:

Part I—Astronomical Parameters

1. Number of Galaxy Stars Suitable for Life (of Spectral Class F, G, K)
2. Number of Suitable Planets in the Habitable Zone per Star (of Spectral Class F, G, K)
3. Fraction of Stable Planets
   (a) Multiple star systems
   (b) Supernovae within 40 ly (light years)
   (c) Gamma-ray bursts within 5000 ly (light years)
   (d) Super flares from their own star
   (e) Transit of gas giants on inner orbits
   (f) Prolonged meteor bombardment
   (g) Instability of the rotation axis
   (h) Absence of the carbon cycle
   (i) Absence of the planetary magnetic field

Part II—Biological Parameters

4. Fraction of Planets Where Life Arises
   (a) The abiological synthesis of biological molecules
   (b) The concentration of the primordial broth
   (c) The formation of lipid bags
   (d) The inclusion of chlorophyll in lipid membranes
   (e) The "proton photopump"
   (f) The formation of nucleic acid filaments
   (g) The catalytic role of RNA
   (h) Determination of roles
   (i) Formation of the cell membrane
   (j) Emergence of the genetic code

5A. Fraction of Planets Where Eukaryotes Arise
    (a) The evolution of an aerobic bacterium
    (b) The host-symbiont encounter
    (c) The formation of pores and the extrusion of extensions
    (d) The "wrapping" of the symbionts and the disappearance of the cell wall of the host
    (e) The "penetration of the symbionts into the cytoplasm"
    (f) The migration of DNA from the genome of the symbiont to that of the host
    (g) The acquisition of the eukaryotic cytoplasmic membrane
    (h) The incorporation into a single coating and phagocytosis
5B. Fraction of Planets in Which Animals (Metzoi) Were Born
    (a) The acquisition of a complex life cycle
    (b) The aggregation of zoospores and the formation of the synzoospore
    (c) The sedentary colony composed of differentiated cells
    (d) The production of collagen
5C. Fraction of Planets Where Technological Civilizations (ETCs) Are Born
    (a) Increase in metazoan size (nervous and vascular system)
    (b) Development of limbs
    (c) Conquest of the mainland
    (d) Differentiation of terrestrial animals
    (e) Acquisition of sociality
    (f) Upright stance and manual dexterity
    (g) Change in diet and brain growth
    (h) Organization of the brain for abstract thought
    (i) Birth of articulated language and technique

PART III—Social Parameters

6. Fraction of Planets Where Life Decides to Communicate
7. Fraction of Duration of the ETCs
    (a) Self-destruction due to evolutionary insufficiency
    (b) Unintentional technological error
    (c) Technological insufficiency to face planetary changes
    (d) Spontaneous involution
    (e) Artificial genetic transition ended on a dead track
    (f) Transition of artificial intelligence ended on a dead track
    (g) Reaching point Ω
    (h) ETCs that overcome the seven challenges and become eternal

The result of this description is the story of life divided into **50** steps: the *trials* that life must overcome to assert itself at various levels of development.

2

The Phases and the Challenges

The **50** steps described, as we will see, have different statistical characteristics that will be examined in due course. The fundamental characteristic underlying all the others is the distinction between a step defined as a *phase* and a step defined as a *challenge*. To immediately grasp the difference between the two, let's use the following analogy: suppose we have a path with various difficulties; for example (Fig. 2.1a, b):

A A maze to exit from (with probability $\mathbf{p_A}$)
B A Tibetan bridge to overcome (with probability $\mathbf{p_B}$)

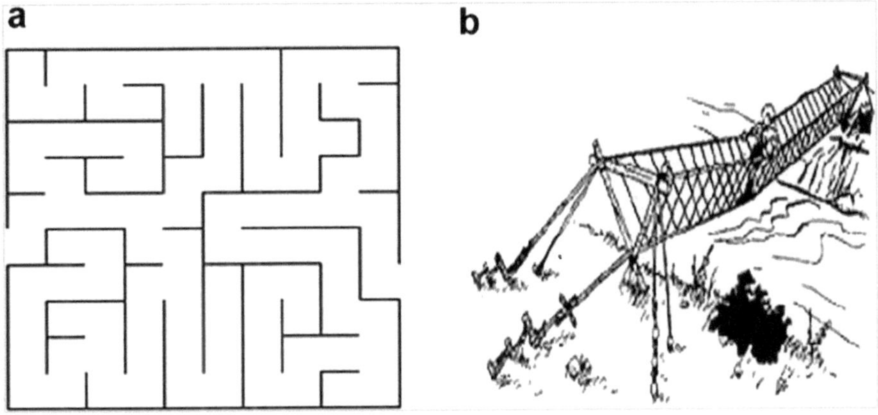

Fig. 2.1 (**a**) A maze to exit from (with probability $\mathbf{p_A}$); (**b**) A Tibetan bridge to overcome (with probability $\mathbf{p_B}$) **(IJA 14/06/2023 Mieli, Valli, Maccone)**

E. Mieli et al., *The Living Galaxy*, https://doi.org/10.1007/978-3-031-67324-5_2

Both of these difficulties have associated the probabilities of overcoming them, $\mathbf{p_A}$ and $\mathbf{p_B}$. Apparently, the statistical nature of these two path steps is the same, but is it really so? No, in fact, *with respect to time*, the two steps behave in opposite ways. That is, the probability $\mathbf{p_A}$ of exiting the maze increases with the time spent in the maze (because over time we understand its intricacies better), while the probability $\mathbf{p_B}$ of overcoming the Tibetan bridge decreases with the time spent on the bridge (because risks increase with time spent in the dangerous situation). The first, borrowing terminology from relativity theory, is *Covariant with Time* and we will define the corresponding step as a *phase*, while the second is *Contravariant with Time* and we will define the corresponding step as a *challenge*.

If $\mathbf{p_A}$ and $\mathbf{p_B}$ do not have an explicit time dependence (this will be true for us only as a first approximation), then their respective transformation laws, going from a reference time interval $\mathbf{\Delta T_0}$ (with probabilities $\mathbf{p_{A0}}$ and $\mathbf{p_{B0}}$) to any $\mathbf{\Delta T} = \mathbf{n} \cdot \mathbf{\Delta T_0}$ (with probabilities $\mathbf{p_A}$ and $\mathbf{p_B}$), will be:

A *Phase* $\mathbf{p_A} = \mathbf{1} - (\mathbf{1} - \mathbf{p_{A0}})^{\mathbf{n}}$ increasing with $\mathbf{n}$
B *Challenge* $\mathbf{p_B} = (\mathbf{p_{B0}})^{\mathbf{n}}$ decreasing with $\mathbf{n}$

It is easy to see that, as $\mathbf{n}$ increases, $\mathbf{p_A}$ approaches $\mathbf{1}$, while $\mathbf{p_B}$ approaches $\mathbf{0}$; in both cases exponentially with respect to $\mathbf{n}$.

If $\mathbf{p_A}$ and $\mathbf{p_B}$ have an explicit dependence on time, then their trend is no longer simply exponential, even if they maintain their nature of phases or challenges.

Part I

The Astronomical Parameters N_s, n_p and f_s

At least in the abstract, the mathematical approach for the first three parameters is less daunting than for the biological ones, the fourth and fifth, because the astronomical events we will now refer to are largely independent of each other without a particular sequence to respect. The same will not be true for the fourth and fifth parameters, as well as the seventh parameter that deals with the fraction of duration of ETCs with respect to **7 Gy**. We now briefly describe the mathematical details related to the Drake equation and the Maccone method:

(a) We will address Drake's first two parameters, $\mathbf{N_s}$ and $\mathbf{n_p}$, without breaking them down into further factors, but directly using what is now strongly emerging from the large mass of experimental data at our disposal
(b) We will instead divide the process of the third parameter, $\mathbf{f_s}$, into **nine** factors representing challenges to overcome, all referred to the single time interval $\mathbf{\Delta T_0}$ = **7 Gy** (billion years) which is the duration of the last stellar population
(c) We will then set, for each single challenge, the input data; namely the minimum and maximum frequencies (fractions of overcoming the challenge) $\mathbf{a_j}$, $\mathbf{b_j}$ of the random variable $\mathbf{x_j}$
(d) Using Maccone's lognormal formula [65] $\mathbf{\Phi(x_0)}$, we will obtain, from these single frequencies, the $\mathbf{<x_0>}$ average frequency of overall planetary survival and the $\boldsymbol{\sigma}_{(X0)}$ standard deviation from the overall average, of the entire process in the period $\mathbf{\Delta T_0}$

(e) From the average frequency we will calculate the minimum and maximum value of the variable $<\mathbf{x_0}>$ with the formula derived from the log-normal (Appendix D):

$$< x_0 >_{min} = \langle x_0 \rangle - \sqrt{3} \cdot \sigma(x_0)$$

$$< x_0 >_{max} = \langle x_0 \rangle + \sqrt{3} \cdot \sigma(x_0)$$

(f) We will extend $<\mathbf{x_0}>_{\mathbf{min}}$ and $<\mathbf{x_0}>_{\mathbf{max}}$, from the period $\mathbf{\Delta T_0}$ of 7 **Gy** to any period $\mathbf{\Delta T}$ (according to the law of transformation of the *challenge*), thus obtaining the final survival frequencies $<\mathbf{x_0}>_{\mathbf{min/max}}(\mathbf{\Delta T})$ of the stable planets as a function of time $\mathbf{\Delta T}$. To do this we set:

$$\begin{cases} m = \dfrac{\Delta T}{\Delta T_0} \\ < \mathbf{X_0} >_{min} (\Delta T) = (< \mathbf{x_0} >_{min} (\Delta T_0))^m \\ < \mathbf{X_0} >_{max} (\Delta T) = (< \mathbf{x_0} >_{max} (\Delta T_0))^m \end{cases}$$

In this way, the new values $<\mathbf{x_0}>_{\mathbf{min/max}}(\mathbf{\Delta T})$ will be referred to the new time $\mathbf{\Delta T}$ which is **m** times the time $\mathbf{\Delta T_0}$. We observe that the expressions above can be written in the more familiar exponential form.

$$\begin{cases} \tau_{(min/max)} \equiv \dfrac{\mathbf{\Delta T_0}}{\ln \dfrac{1}{< \mathbf{X_0} >_{min/max}}} \\ < \mathbf{X_0} >_{min} (\mathbf{\Delta T}) = \exp\left(-\dfrac{\mathbf{\Delta T}}{\tau_{min}} \right) \\ < \mathbf{X_0} >_{max} (\mathbf{\Delta T}) = \exp\left(-\dfrac{\mathbf{\Delta T}}{\tau_{max}} \right) \end{cases}$$

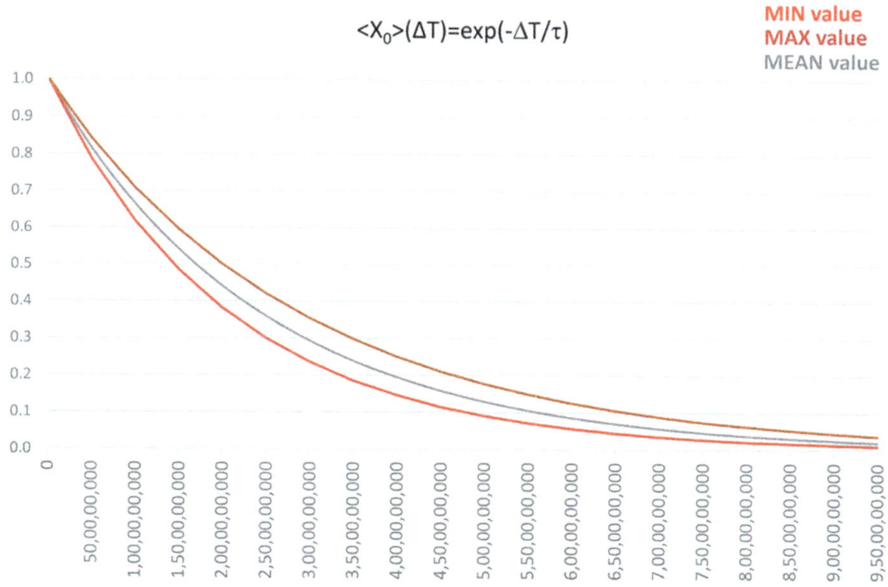

Fig. I.1 Expected trend from the calculation of **<x1>min/max** as a function of **ΔT** with **tmed~2.5 Gy (IJA 14/06/2023 Mieli, Valli, Maccone)**

The Fig. I.1 shown represents the expected trend from the calculation of $\mathbf{<X_0>_{min/max}}$ as a function of $\mathbf{\Delta T}$ (an example was taken, $\mathbf{t_{med}\sim 2.5}$ **Gy**). As you can see, the three values $\mathbf{<X_0>_{min}}$, $\mathbf{<X_0>_{max}}$ and $\mathbf{<X_0>_{med}}$ have an exponentially decreasing trend as a function of the reference duration $\mathbf{\Delta T}$. This result will be useful later for determining the number of planets suitable for different levels of life development.

3

1st Drake: The Stellar Numerosity of the Galactic Disk for K, G and F Spectral Class Stars

Let's start with the 1st Drake parameter, the stellar numerosity of the galaxy, but also assign to the first parameter the scope of considering only the stars today deemed suitable for the development of life, namely those of spectral class **K**, **G** and **F** of the galactic disk [26] (Fig. 3.1).

The reason for this choice, in addition to balancing the computational effort between the parameters, as anticipated in the introduction, also has a mathematical advantage: if, for example, we now considered all the main sequence stars of the Milky Way, we would have to say they are about between $\mathbf{2 \cdot 10^{11}}$ and $\mathbf{4 \cdot 10^{11}}$, or on average, $\mathbf{3 \cdot 10^{11}}$ with a very large deviation of $\mathbf{10^{11}}$; therefore, even if the underestimation/overestimation of the average value would then be absorbed by the second parameter, the deviation from the average value would not, resulting in an unnecessarily large impact on the final calculation of the Drake equation.

We must note that the exclusion of the numerous spectral class **M** (red dwarfs with masses between **0.08** and $\mathbf{0.45\ M_\odot}$ solar masses) from this computation depends on the current belief that such solar systems are highly unstable from the point of view of the star's nuclear activity. Due to their slow stellar evolution caused by their small mass, these stars present frequent and lethal flares (ejection of stellar material in the vicinity of the star) for the entire first part of the planets' lifetimes [16]. The planets, in turn, for these stars have synchronous rotations (they always face the same side towards the star) due to the strong tidal forces in the habitable zone. These two factors, especially the first, make this type of star unsuitable for hosting life for a long time, despite the recent discoveries of wonderful planetary systems, apparently habitable, around such stars.

E. Mieli et al., *The Living Galaxy*, https://doi.org/10.1007/978-3-031-67324-5_3

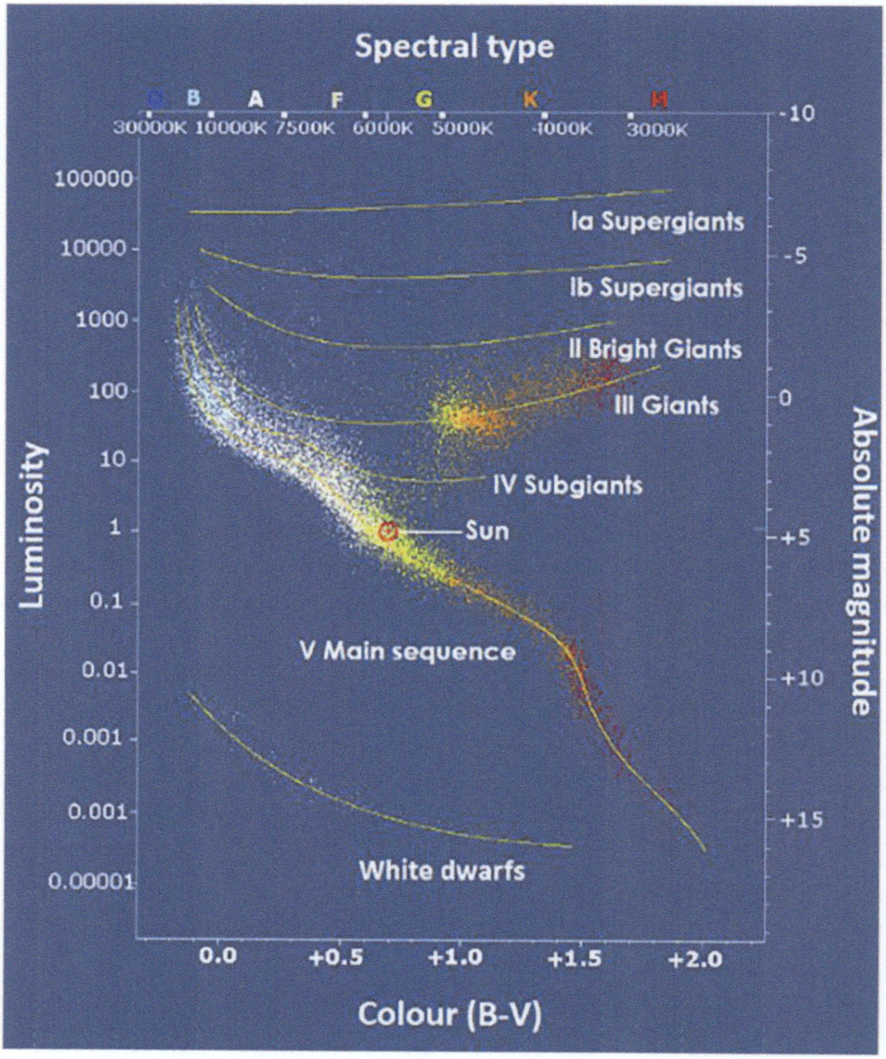

Fig. 3.1 Hertzsprung-Russell diagram (original image by Richard Powell modified by Helen Klus)

The same reasoning has been applied to the galactic center (*bulge*), considered too crowded and therefore subject to frequent violent phenomena, such as supernovae or gamma-ray bursts, lethal for the development of life [69].

Based on these considerations, we analyze the data of Table 3.1, which reports all stellar types divided by areas of the galaxy. It is clear that we are only interested in **F**, **G** and **K** class stars belonging to the galactic disk: the sum of these three sets is equal to $\mathbf{1.10 \cdot 10^{10}}$ stars. Despite the presence of three significant figures in the mantissa, we prudentially assign an uncertainty of **±10%** to the 1st Drake parameter, that is:

Table 3.1 1st Drake: stellar numerosity of the Milky Way divided by location and spectral type

Spectral class	Mass ($M_\odot$)	% stars disk	Core stars $2.40 \cdot 10^{10}$ $M_\odot$	Disk stars $3.79 \cdot 10^{10}$ $M_\odot$	Halo stars $2.40 \cdot 10^{8}$ $M_\odot$	Primary sequence stars $6.43 \cdot 10^{10}$ $M_\odot$
O	≥16	10^{-5} %	$8.57 \cdot 10^{1}$	$1.35 \cdot 10^{2}$	([a])	$2.21 \cdot 10^{2}$
B	2.1–16	0.13%	$2.99 \cdot 10^{6}$	$4.71 \cdot 10^{6}$		$7.70 \cdot 10^{6}$
A	1.4–2.1	0.60%	$6.40 \cdot 10^{7}$	$1.01 \cdot 10^{8}$		$1.65 \cdot 10^{8}$
F	1.04–1.4	3.00%	$5.43 \cdot 10^{8}$	$8.58 \cdot 10^{8}$		$14.00 \cdot 10^{9}$
G	0.8–1.04	7.60%	$1.97 \cdot 10^{9}$	$3.11 \cdot 10^{9}$		$5.09 \cdot 10^{9}$
K	0.45–0.8	12.10%	$4.47 \cdot 10^{9}$	$7.06 \cdot 10^{9}$	$1.71 \cdot 10^{9}$	$1.32 \cdot 10^{10}$
M	0.08–0.45	76.45%	$6.44 \cdot 10^{10}$	$1.02 \cdot 10^{11}$	$2.27 \cdot 10^{9}$	$1.68 \cdot 10^{11}$
Total predicted stars in the Milky Way			$7.14 \cdot 10^{10}$	$1.13 \cdot 10^{11}$	$3.97 \cdot 10^{9}$	$1.88 \cdot 10^{11}$

[a]The production of new stars of these spectral classes in the halo ceased **10 billion** years ago and only small Population II stars are still in the main sequence. The others have gone extinct (data taken from the site https://ilpoliedrico.com/2015/12/quante-stelle-ci-sono-nella-via-lattea.html)

Drake 1

$N_{s\,min}$	$N_{s\,max}$
$1.0 \cdot 10^{10}$	$1.2 \cdot 10^{10}$

4

2nd Drake: Number of Planets per Star, Suitable for Life in the Habitable Zone (Spectral Class F, G and K)

The second parameter will also be directly estimated from the literature. In this case, the 2020 work of Michelle Kunimoto and Jaymie M. Matthews, which is based on an independent catalog of extrasolar planets compiled from about **200,000** stars, directly provides the result we need [56] (Fig. 4.1).

For planets with sizes **0.75–1.5 $R_\oplus$** (Earth radii) in a conservatively defined habitable zone (**0.99–1.70 AU**, astronomical units) around G-type stars,

Fig. 4.1 Showing in green the habitability zone or "Goldilocks zone"

E. Mieli et al., *The Living Galaxy*, https://doi.org/10.1007/978-3-031-67324-5_4

Kunimoto defines an average value of **0.18** planets per star with an uncertainty of **10%.** That is:

Drake 2

$n_{p\ min}$	$n_{p\ max}$
0.16	0.20

5

3rd Drake: Fraction of Stable Planets for 7 Gy (Duration of the Stellar Population)

With the third parameter, we finally enter the calculation by dividing the process into several challenges (in this case **9**) that the planet survives with a certain minimum and maximum probability for each. The use of Maccone's formula [67] collects all the input data and provides the minimum and maximum values of the overall 3rd Drake parameter.

The nine challenges, which represent the astronomical dangers to which the planet is subjected, are as follows:

1. Multiple star systems
2. Supernovae less than 40 ly away
3. Gamma bursts less than 5000 ly away
4. Super-flares of its own star
5. Transit of gas giants on inner orbits
6. Prolonged meteor bombardment
7. Instability of the rotation axis
8. Absence of the carbon cycle
9. Absence of the magnetic field

5.1 Challenge 1: Multiple Star Systems

Until a few years ago, the possibility of planets existing around multiple star systems was considered residual. Today we have concrete evidence, the most famous of which is the exoplanet *Proxima Centauri b*—a planet not belonging to our solar system and also the nearest known exoplanet to us. Proxima

E. Mieli et al., *The Living Galaxy*, https://doi.org/10.1007/978-3-031-67324-5_5

Centauri, along with Alpha Centauri A and B, is even a triple star system: Alpha Centauri A and B orbit around their common center of mass at a close distance, while Proxima Centauri (a red dwarf of spectral class M5Ve, which we will only take as an example here) orbits this pair at a much greater distance. The *Proxima Centauri b* planetary system is defined as a **Type S planetary system**, meaning the planets orbit around an essentially isolated star because it is very far from its companion(s) (in this case, the Alpha Centauri pair). Systems of this type are quite common and do not preclude the formation of living forms, as each S-type system has its own habitable zone scarcely influenced by the distant orbiting star.

On the contrary, the opposite case is that of the **type P planetary systems**, called ***circumbinary***, composed of a pair of stars orbiting closely together with a planetary system orbiting at a much greater distance around the pair (like the fictional planet Tatooine in Star Wars Episode IV from 1977). Such systems have an ovoid habitable zone that follows the motion of the stellar pair, making it ill-suited for stable planetary orbits. Therefore, we can reasonably exclude such systems from those capable of hosting life forms [100] (Fig. 5.1).

In conclusion, we must exclude the number of multiple star systems with tight orbital distances (which could potentially form circumbinary planetary systems) from the total of **F**, **G**, and **K** spectral class stars. As indicated by Charles J. Lada in the article "Stellar Multiplicity and the IMF: Most Stars

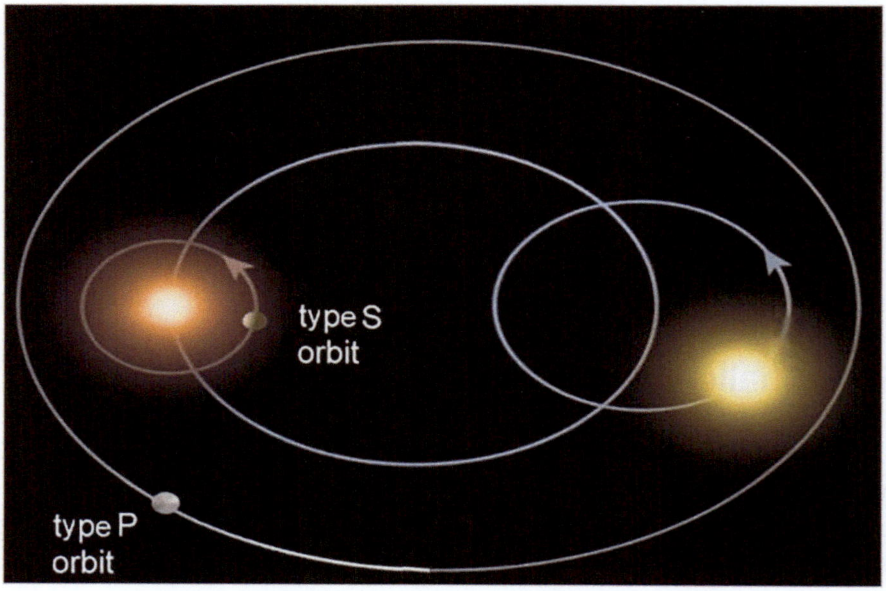

Fig. 5.1 **P** and **S** planetary systems

Are Single", for stars with masses similar to the Sun, the study finds **56%** are single stars and **44%** are double or multiple star systems. From this **44%** of multiple stars, we must estimate what fraction have orbits that are not widely detached and could produce circumbinary systems—prudently between **10** and **30%**. Therefore, the fraction of **F**, **G**, and **K** stars suitable for life is likely between **70%** (minimum frequency $\mathbf{a_1}$) and **90%** (maximum frequency $\mathbf{b_1}$).

Challenge 1

a_1	b_1
0.7	0.9

5.2 Challenge 2: Supernovae Less Than 40 ly Away (Safety Distance)

We can quickly estimate the risk factor due to supernovae through the following considerations:

(A) From indirect evaluations, we estimate there are about three supernova explosions per century in the Milky Way.
(B) Over the last 7 **Gy** of stellar population, the Milky Way has therefore witnessed around **200 million** supernova explosions, or roughly one for every **1000** stars.

Assuming these **1000** stars are distributed uniformly in volume, the average distance from any given star to a supernova is about $\mathbf{1000^{1/3} = 10}$ times the average distance between stars in the galactic disk, which is around **5 ly** (light-years). Therefore, the average distance of a supernova from another star is approximately **50 ly** (Fig. 5.2),

Obviously, this is only an average value with occasional deviations. It's estimated that a supernova occurs within **50 ly** of Earth every **250 million** years, potentially compromising part of the ozone layer with partially lethal effects on the biosphere [14]. For these reasons, we have assigned a **70%** probability, with a **10%** deviation, of a planetary system escaping danger due to a supernova explosion. In other words:

Challenge 2

a_2	b_2
0.6	0.8

Fig. 5.2 Explosion of a supernova Powered by OpenAI

5.3 Challenge 3: Gamma Bursts Less Than 5000 ly Away (Safety Distance)

The phenomenon of gamma-ray bursts has an extensive literature dedicated to investigating their causes and potential effects on planets struck by these immense events. It is estimated that they emit an energy equal to **10^{44}–10^{45} J** (Joules) in just a few seconds, equivalent to the entire energy output of the sun over its lifetime [116] (Fig. 5.3).

Being extremely powerful phenomena, gamma-ray bursts are detectable even at the farthest observable limits of the universe. This characteristic aids our task because, for this reason, without delving into the details of their

Fig. 5.3 Gamma-ray burst formed by two highly collimated jets moving in opposite directions (**European Southern Observatory**)

nature and emission modes (isotropic or collimated beams), we can directly make a statistical estimate from the observed number of such phenomena recorded on Earth.

(A) About one gamma-ray burst is detected per day from the observable universe in the direction of Earth. Therefore, over the last 7 **Gy** stellar population, there have been about $\mathbf{2.5 \cdot 10^{12}}$ gamma-ray bursts in our direction.
(B) There are approximately $\mathbf{10^{23}}$ stars in the observable universe.
(C) Therefore, over the last stellar population there has been one gamma-ray burst for every $\mathbf{4 \cdot 10^{10}}$ stars.
(D) The volume defined by the safety distance of $\mathbf{5000\ ly = 5 \cdot 10^3\ ly}$ from the nearest gamma-ray burst is

$$\left(5 \cdot 10^3\right)^3 = 125 \cdot 10^9\ \text{ly}^3$$

(E) The density of the galactic disk is approximately $\mathbf{5^3 = 125\ ly^3/star}$ (1 star per 125 cubic light-years).
(F) Therefore, within the 5000 light-year safety volume, there are $\mathbf{10^9}$ stars.
(G) This corresponds to an average of $\mathbf{10^9/4 \cdot 10^{10} = 2.5 \cdot 10^{-2}}$ gamma-ray bursts within the safety volume over the last stellar population—a very low value. However, Melott hypothesized in 2003 that the Ordovician-Silurian mass extinction was due to a gamma-ray burst.

As in the previous case, we cautiously place ourselves slightly above this risk margin, let's say at $\mathbf{5 \cdot 10^{-2}}$ with a deviation of $\mathbf{5 \cdot 10^{-2}}$ to account for the great uncertainty around these phenomena. The complementary value indicating the probability of avoiding the gamma-ray burst risk is therefore:

Challenge 3

a_3	b_3
0.9	1.0

5.4 Challenge 4: Super-Flares from One's Own Star

We now move to consider the dangers coming directly from one's own planetary system. A *stellar super-flare* is commonly defined as a violent eruption of matter exploding from a star's surface, with energy equivalent to a million times or more than that of typical solar flares (Fig. 5.4).

One characteristic of solar-type stars that have exhibited super-flares is that they have faster rotation and higher magnetic activity than the Sun. It has been hypothesized that such explosions are produced by the interaction of the stellar magnetic field with that of a close-orbiting hot Jupiter (giant gaseous planet); however, this theory lacks confirmation even after searching for hot Jupiters around stars that have presented super-flare phenomena. In a 2012 study by Maehara [67], **83,000** sun-like stars were analyzed using data from the Kepler space telescope, finding **365** super-flares originating from **148** stars, with an average duration of **12** h, over **120** days—at first glance, a statistically very high occurrence. But we ignore how these phenomena are distributed on the totality of the stars like the sun during all **7 Gy** of the star population; we only know that in **120** days **148** stars of **83,000** had a super-flare, but we don't know if:

(a) Always the same **148** stars will present super-flares
(b) The super-flares are distributed evenly on all stars

The case (**a**) is probably closer to the truth. Let's see why.

The 2012 work by Maehara [67] provides us with the following elements on the number of superflares affecting solar-type stars, i.e., spectral class **G**:

N = **83,000** solar-type stars under examination
T = **120 d** days of observation

Fig. 5.4 Super flare Powered by OpenAI

n = **363**	detected superflares	
m = **148**	stars affected by superflares	
t = **12 h**	average duration of superflares in hours	
p = **?**	probability of superflares in time **T**	unknown to be found
α = **?**	fraction of stars subject to superflares	unknown to be found

We made two crucial hypotheses to follow this reasoning:

1 The probability p of superflare is constant and does not explicitly depend on time or on the occurrence of other superflares on the star in question

2 The stars are divided into only two categories **A** and **B**. Those that exhibit superflares (**A**: fraction α) and those that do not exhibit the phenomenon (**B**: fraction $1 - \alpha$)

We will have:

$$\begin{cases} \alpha \cdot N \cdot p = n \\ \alpha \cdot N \cdot (p^2 + p^3 + \ldots) = (n - m) \\ p^2 + p^3 + \ldots = \left(\dfrac{p^2}{1-p} \right) \end{cases}$$

- The first of the three equations counts the total number of superflares based on the probability p
- The second equation counts double, triple, etc. superflares, always based on the probability p, knowing that only m = 148 stars were affected by the phenomenon
- The third equation is simply the sum of the series of powers of p starting from the second

Solving the system, we obtain:

$$\begin{cases} p = \dfrac{n-m}{2n-m} \cong 0,37 \\ \alpha = \dfrac{n \cdot (2n-m)}{N \cdot (n-m)} \cong 1,2 \cdot 10^{-2} \end{cases}$$

Where the value we are interested in is α = **1.2%**, which tells us that the fraction α of stars affected by superflares is very low, despite the high probability p = **37%** of superflares occurring on that fraction of stars. This would lead us to a probability of overcoming this challenge close to **99%**.

However, we observe that, even if the two conditions **1** and **2** were satisfied, we cannot say that only **1.2%** of stars are at risk because our observation period **T**, equal to **120** days, is practically instantaneous compared to the stellar generation period (7 **Gy**) and does not allow us to understand if, during their evolution, stars can pass from state **A** to state **B**. For this reason, we must remain cautious in evaluating the probability of overcoming this challenge by assigning a probability between **0.5** and **0.7**:

For this reason, the absence of associated mass extinctions suggests that no super-flare from the Sun has occurred in the past. We therefore associate these phenomena with an intermediate risk level and a consequent survival probability slightly above **50%**:

Challenge 4

a_4	b_4
0.5	0.7

5.5 Challenge 5: Transit of Gas Giants on Inner Orbits

Until the discovery of the first exoplanets, mathematical models of planetary evolution enjoyed relative tranquility, being able to refer to the only known example—our solar system. This scenario is as orderly as a museum hall, with rocky and massive planets in the inner belt (but still not too close to the sun) and gas giants in the outer belt; there is even a transition zone, the asteroid belt, which, as predicted, could not aggregate into a planet due to the immense gravity of nearby Jupiter (Fig. 5.5).

Then we began receiving the first data on observed exoplanets and realized that our orderly solar system is an exception, and instead most planetary systems are totally chaotic—like a teenager's room: gas giants at ridiculous distances from their stars, **0.05 AU** or even less; gigantic super Earths with tens of Earth masses positioned haphazardly in relation to other planetary masses. In short, we had to revise everything and correct our models taking into account the new data. It seems planetary systems are intrinsically unstable and only rarely remain in their *natural positions*, if such positions exist, as happened in our system.

In this situation, however, one thing is certain: the presence of a nearby gas giant disturbs, with its large mass, if not the formation of other planets (as in the asteroid belt case) certainly the trajectory of transiting meteoritic objects that would be dangerously attracted to inner orbits close to rocky planets in the habitable zone.

Through exoplanet observations we can affirm that such an event is not rare [47]. However, we should not fall into the trap of considering the statistics collected from exoplanets as directly relatable to the galactic planetary reality, since some types of planets, the hot Jupiters indeed, are those most easily

Fig. 5.5 Hot Jovian planet in tight orbit around the star (**NASA, ESA and A. Schaller (for STScI)**)

observable both with the transit method and radial velocity method, which are most commonly used. For this reason, we will give the transit of gas giants on inner orbits an average probability of **20%** with a deviation of **10%**; which, compared to the complementary probability of life surviving on a planet in the habitable zone, translates into the following values:

Challenge 5

a_5	b_5
0.7	0.9

5.6 Challenge 6: Prolonged Meteoric Bombardment

There would be no reason to think that the typical meteoric bombardment accompanying the formation of rocky planets should prolong much beyond the initial phases of the planet's life [37], especially if we have already excluded the nearby presence of gas giants that can attract objects of various sizes from outer asteroid belts (for the solar system, the Kuiper belt and Oort cloud) (Fig. 5.6).

However, as we have seen, the solar system is not a typical example of a planetary system, so we cannot exclude that, for example, a particular arrangement of super Earths in or near the habitable zone can determine gravitational resonance phenomena with possible asteroid belts. We do not have experimental evidence in this regard, however, we want to give a statistical weight, even if low, to this potential risk:

Challenge 6

a_6	b_6
0.8	1.0

Fig. 5.6 Impact of meteorites on a rocky planet (**NASA's Goddard Space Flight Center Conceptual Image Lab**)

5.7 Challenge 7: Instability of the Rotation Axis

Normally, the rotational axes of planets do not have a stable inclination, but vary chaotically due to interactions between their orbits. Typical variability periods are on the order of a few million years (see Mars) and the oscillation intervals are very large (**60–90°**). Earth does not have this characteristic because its large satellite, the Moon, acts as a natural stabilizer of the axial inclination, except for minor variations [101] (Fig. 5.7).

But how important is this characteristic for the development of life? Given the massive climate impacts that would result from a chaotic inclination of a planet's axis, it is plausible that such a situation, would not completely prevent life from arising, but would constrain its evolution to the most resilient and primitive forms.

Even though data related to the presence of *exomoons* (satellites of exoplanets) in the planetary systems being discovered daily are currently completely absent, in this case it is not wrong to imagine that, predominantly, large moons are a characteristic of gas giants and not smaller rocky planets. Therefore, a situation like Earth's is likely rare. As a result, the following survival values emerge for this challenge:

Challenge 7

a_7	b_7
0.1	0.3

Fig. 5.7 Different inclinations with respect to the plane of revolution (which is assumed to be horizontal in the figure) of the rotation axes of the planets

5.8 Challenge 8: Absence of the Carbon Cycle

The stability of temperature is not maintained solely by a stable and not excessive axial tilt, but also by appropriate mechanisms for eliminating and restoring greenhouse gases in the planet's atmosphere, particularly $\mathbf{CO_2}$. On Earth, an effective natural thermostat is the so-called *carbon cycle*—the dynamic interchange between the geosphere, hydrosphere, biosphere and atmosphere through chemical, physical, geological and biological processes [55]. In summary, the Earth's carbon cycle thermostat mechanism works as follows (Fig. 5.8):

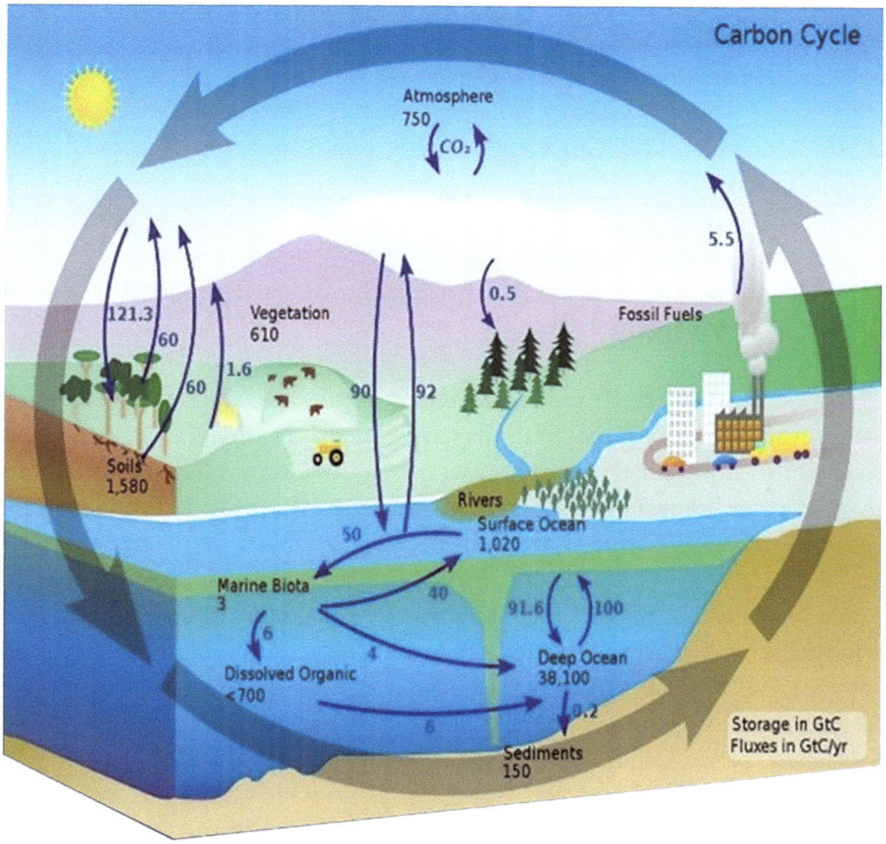

Fig. 5.8 Diagram of the carbon cycle in the terrestrial environment

Carbon Capture Cycle as CO_2:

(a) The planet is initially in the habitable zone with liquid water and average temperatures around **20 °C**
(b) Volcanism (or animal activity or other sources) enriches the atmosphere with CO_2, causing an initial imbalance
(c) The CO_2 increases the planet's temperature
(d) The thermal increase intensifies evaporation of seas and resulting rainfall
(e) The rains absorb the excess CO_2 and return it to the oceans
(f) The carbon present in the CO_2 precipitates on the seabed
(g) Subduction phenomena (sliding of one tectonic plate under another) return the carbon to the geosphere

Carbon Restoration Cycle as CO_2:

(h) The atmosphere poor in CO_2 does not retain heat and the planet cools (glaciation)
(i) Volcanism restores CO_2 without causing further rains as long as temperature remains low
(j) Temperature increases due to the greenhouse gas effect and returns to starting levels (point **a**)

It's thought that Earth's carbon cycle has been active since the planet's formation, though the specific methods have varied over time. Clearly, the main driver of this mechanism is geothermal activity working in combination with the other three components. Therefore, the essential condition for a thermally stabilized planet is **active plate tectonics**—movement of plates relative to each other.

Paradoxically, the planets that have this characteristic are moderately-sized rocky ones, not the smallest rocky planets. The reason is that smaller planets—half Earth's size, for example—have greater thermal dissipation and therefore exhaust their internal heat supply earlier. At that point, the planet cools, the crust solidifies uniformly and geological activity stops, along with the carbon cycle. This is what happened to Mars which, in addition to insufficient gravity to retain a dense atmosphere, lacks a carbon cycle to maintain temperature in the habitable zone.

On the contrary, the recently discovered so-called *super Earths*—a rich class of rocky planets equal to or larger than Earth in size—would be excellent candidates in this respect, likely having more active plate tectonics than Earth.

Recently, a theoretical model has been developed that predicts whether a carbon cycle is present on exoplanets, given the mass, core size and amount of **CO_2**. In any case, super Earths are very common, ensuring that the risk of not finding geologically active planets is quite low. So let's assume:

Challenge 8

a_8	b_8
0.7	0.9

5.9 Challenge 9: Absence of the Planetary Magnetic Field

There is another problematic and still little-studied aspect: the analysis of the deep interiors of planets and their magnetic fields. We know that the magnetic field shields a planet from stellar winds of charged particles, and is therefore as necessary for life as the ozone layer is for blocking ultraviolet radiation (Fig. 5.9).

In the case of super Earths, a 2021 study conducted by the Earth and Planets Laboratory at the Carnegie Institution for Science [28], a private scientific institute based in Washington D.C., simulated extreme pressures

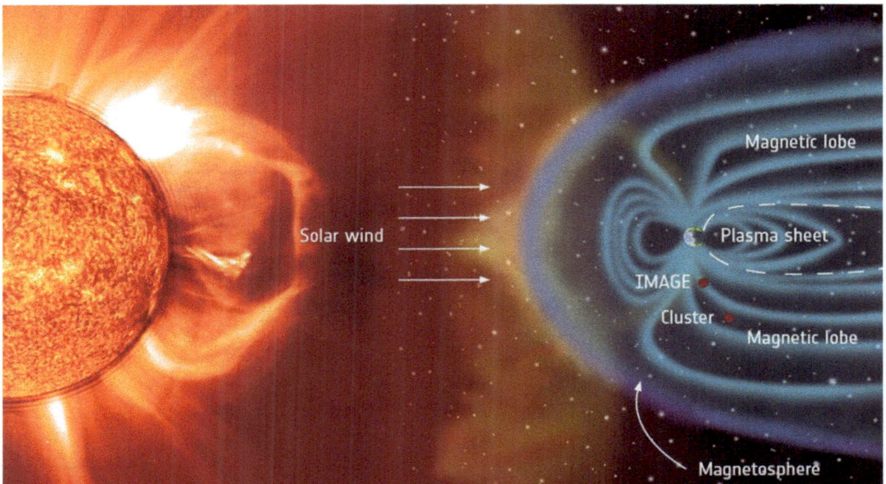

Fig. 5.9 Deviation of charged stellar particles by the planetary magnetic field **(ESA/NASA/SOHO/LASCO/EIT)**

exerted on magnesium silicate to better understand the interior dynamics, particularly of the mantle, of rocky exoplanets most similar to Earth. The goal was to discover if such planets can generate a magnetic field similar to Earth's and therefore host life.

The conclusions are nuanced: in some geological scenarios, super Earths could indeed generate an Earth-like *geodynamo* early in their evolution, but then lose it over billions of years as cooling slows down the dynamo action. A resurgence of magnetic activity could potentially be triggered by the movement of lighter elements during inner core crystallization.

Given this uncertain situation, we think it reasonable to assign a 50% probability to this challenge of super Earths maintaining magnetic fields conducive for life over long timescales:

Challenge 9

a_9	b_9
0.4	0.6

5.10 The Calculation of the Third Parameter from the Nine Challenges

In conclusion, we have obtained for the 3rd Drake parameter, over the time span of the entire stellar population 7 **billion** years (Table 5.1, Fig. 5.10):

Table 5.1 3rd Drake: input data insertion a_j, b_j and ΔT_0 in the lognormal formula. The light gray shows the intermediate calculations of the logarithmic mean and logarithmic variance

	a_j	b_j	μ_j	σ^2_j	ΔT_0	μ	σ^2	$<X_0>$	$f_{s\,min}$	$f_{s\,max}$
1	0.70	0.90	−0.2258	0.0052	7Gy	−4.07	0.15	1.84%	0.58%	3.11%
2	0.60	0.80	−0.3601	0.0069						
3	0.90	1.00	−0.0518	0.0009						
4	0.50	0.70	−0.5155	0.0094						
5	0.70	0.90	−0.2258	0.0052						
6	0.80	1.00	−0.1074	0.0041						
7	0.10	0.30	−1.6547	0.0948						
8	0.70	0.90	−0.2258	0.0052						
9	0.40	0.60	−0.6999	0.0136						

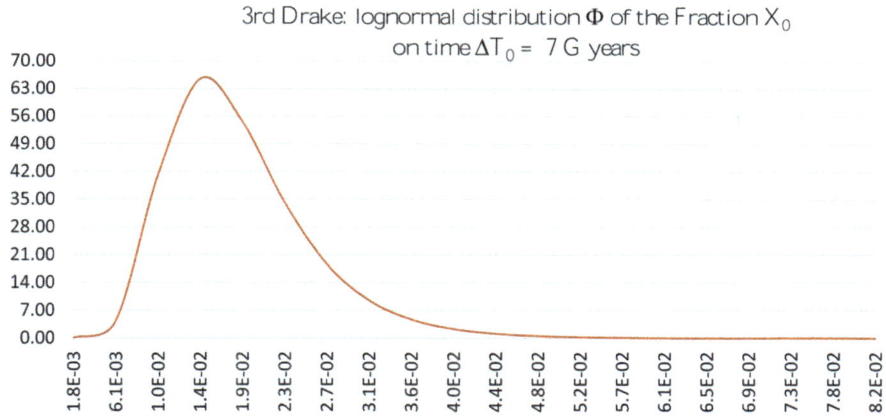

Fig. 5.10 3rd Drake: trend of the lognormal curve for the nine input values. The average value found for the third parameter is therefore about **1.9%** (between **0.6** and **3.1%**) (**IJA 14/06/2023 Mieli, Valli, Maccone**)

Drake 3

$f_{s\ min}$	$f_{s\ max}$
$5.8 \cdot 10^{-3}$	$3.1 \cdot 10^{-2}$

6

Considerations on the First Three Parameters

At this point we can insert precise values into the formula for the final survival frequencies $\mathbf{f}_{s(min/max)}(\Delta\mathbf{T})$ of stable planets as a function of stability time $\Delta\mathbf{T}$.

Or:

$$\begin{cases} \tau_{(min/max)} \equiv \dfrac{\Delta \mathbf{T}_0}{\ln \dfrac{1}{f_{s\,min/max}}} \\ \mathbf{f}_{s\,min}(\Delta \mathrm{T}) = \exp\left(-\dfrac{\Delta \mathbf{T}}{\tau_{min}}\right) \\ \mathbf{f}_{s\,max}(\Delta \mathrm{T}) = \exp\left(-\dfrac{\Delta \mathbf{T}}{\tau_{max}}\right) \end{cases}$$

Which in our case will be (also adding $\boldsymbol{\tau}_{med}$):

$$\tau_{min} = 1{,}359{,}230{,}076 \text{ years}$$
$$\tau_{max} = 2{,}016{,}147{,}681 \text{ years}$$
$$\tau_{med} = 1{,}752{,}672{,}805 \text{ years}$$

The Fig. 6.1a shows the trend of the probability of planetary stability as a function of time $\Delta\mathbf{T}$ throughout the arc of the stellar population of 7 **Gy**. The Fig. 6.1b, instead, is nothing more than the probability of habitability at the present time obtained by multiplying the previous value by $\Delta\mathbf{T}/7\mathbf{Gy}$. These

E. Mieli et al., *The Living Galaxy*, https://doi.org/10.1007/978-3-031-67324-5_6

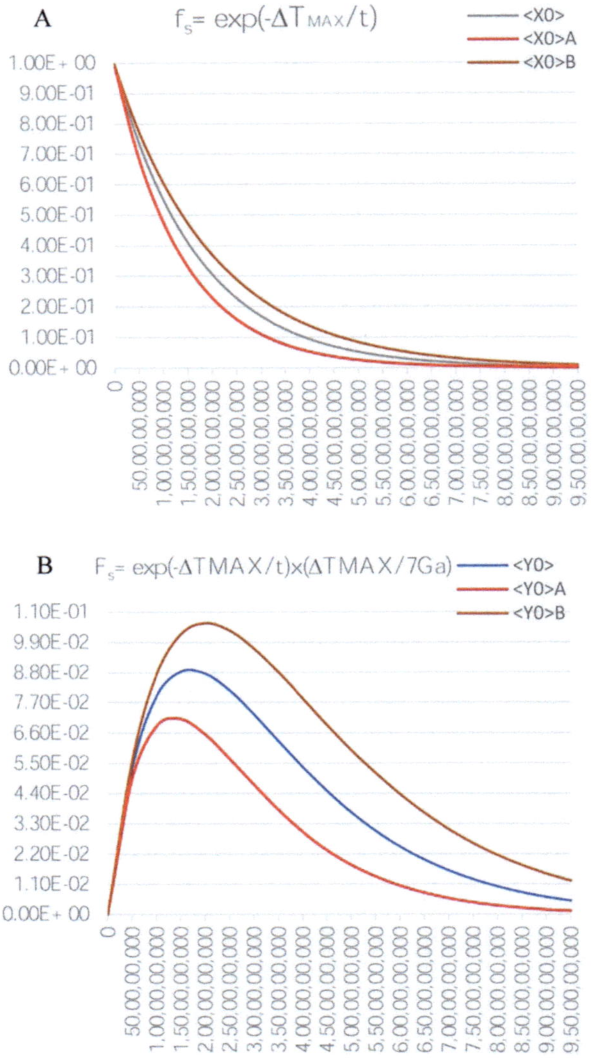

Fig. 6.1 (a) 3rd Drake: calculated trend of the habitability probability $\mathbf{f_{s\ min/max}}$ as a function of **ΔT** with τ_{med} **= 1.75 Gy** calculated **(IJA 14/06/2023 Mieli, Valli, Maccone)**. **(b)** 3rd Drake: current habitability probability $\mathbf{F_{s\ min\ /max}} \equiv \mathbf{f_{s\ min\ /max}} \cdot$ **ΔT/7Gy (IJA 14/06/2023 Mieli, Valli, Maccone)**

trends, together with the remaining Drake parameters, will be useful in our subsequent discussion to count the planets at different levels of development of life and of extraterrestrial civilizations.

For the 3rd Drake parameter we found that, despite the significant screening carried out with the first two parameters for stars and planets suitable for the development of life, the curve of planetary stability $\mathbf{f_s(\Delta T)}$ of the third parameter drops quickly to zero. In the case of systems of stable duration of at least **7 Gy**, such as to allow with reasonable certainty the development of intelligent civilizations, the value of the probability $\mathbf{f_s}$ drops to **1.84%** which is a decidedly modest value.

Part II

Drake's Biological Parameters: f_l and f_i

It goes without saying that the mathematical approach to the so-called *biological* parameters, the fourth and fifth, will necessarily be more complex than the first three parameters. The reason is that biological and evolutionary processes almost always must follow a rigid sequence under delicate environmental conditions (Fig. II.1).

To statistically simulate such a scenario by dividing it into *phases* (we will call them this instead of *challenges*, as done previously, because they are now

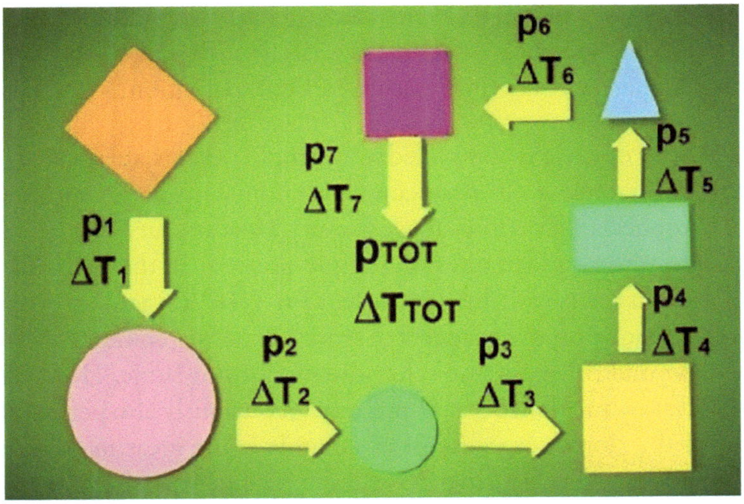

Fig. II.1 Fourth sequence of phases **j** that constitute the biological parameters with associated times **ΔTj** and probabilities **pj**

covariant with time, meaning they increase with the duration of each phase) using Maccone's lognormal distribution, we will proceed as follows:

(A) We will divide the phenomenon (with the associated parameter, e.g. the fourth parameter for the appearance of life) into **n** appropriate stages.
(B) We will set observation time intervals ΔT_j corresponding to each stage **j**, and the related maximum and minimum frequencies a_j and b_j (relative to the random variable x_j) for realizing that stage.
(C) We will define maximum time intervals ΔT_{0j} of ***microcatastrophe*** for each stage **j**, i.e. the times beyond which environmental conditions change and effectively prevent realization of that stage.
(D) We will transform all frequencies a_j and b_j into their corresponding A_j and B_j (random variable X_j) related to the microcatastrophe times ΔT_{0j}.
(E) We will add up all the times ΔT_{0j} obtaining the order of magnitude of the duration of the entire process of the **n** stages equal to ΔT_0.
(F) We will calculate the mean $\langle X_0 \rangle$ and standard deviation $\sigma(X_0)$ of the phenomenon formed by the **n** temporally homogeneous stages, through Maccone's lognormal distribution $\Phi(X_0)$ referred to time ΔT_0, consequently obtaining $\langle X_0 \rangle_{min}$ and $\langle X_0 \rangle_{max}$.
(G) We will define a maximum ***macrocatastrophe*** time interval ΔT for all stages, i.e. the observation time beyond which large planetary events occur that prevent realization of the entire n-stage sequence (e.g. mass extinction events).
(H) We will transform $\langle X_0 \rangle_{min}$ and $\langle X_0 \rangle_{max}$ into their values f_{min} and f_{max}, related to the macrocatastrophe time ΔT, which will be the fourth and fifth parameters of the final calculated Drake equation.

Once a phase or cycle is completed, catastrophic events are no longer lethal for that phase or cycle itself. An example is anomalous storm surges (case of microcatastrophes) that periodically, e.g. every **10** years, hit quiet lagoon environments, causing a sudden dilution of lipid pockets. In this case, the pocket concentration drops below the limit level of $n = 10^8/m^3$ and, as we will see, this blocks Ageno's proposed process of remixing pocket contents, delaying its realization beyond **30** years until after the next storm surge. However, if the reaction occurs before the **10-year** limit time, it obviously also persists through the onset of the storm surge. Another example is the great mass extinctions (case of macrocatastrophes) that have reduced, but not eliminated, life on the planet.

Therefore, the different phases exhibit a certain resilience to traumas following their completion.

7

4th Drake: The Transition from Non-living to Living

The origin of life and its possible occurrence beyond our planet constitute one of the most exciting challenges facing modern science. It is not merely a matter of formulating a consistent hypothesis on the transition from non-living to living matter, which is itself a profound problem. We must also assess the possibility that such a process can occur elsewhere in the Universe and with what probability.

Many hypotheses and theories on the origin of life have been proposed in the past [33, 52]. However, the currently most well-known and accredited is that put forth by British biochemist Nick Lane [59]. He began with the idea that all living beings possess six specific characteristics: (1) a carbon source, (2) an energy source (the same or different from the first) to power metabolic processes (i.e. all growth and functioning of the organism), (3) catalysts (specific molecules with defined tasks) suitable for promoting chemical reactions, (4) the ability to expel waste (i.e. harmful byproducts of cellular processes), (5) compartmentalization (separation between interior and exterior environments), and (6) the presence of heritable material. It is easy to recognize in these activities the basic functions of a simple prokaryotic cell (without a nucleus or organelles) such as a bacterium.

Lane situates this process in alkaline hydrothermal vents [15], where, by exploiting differences in pH (i.e. acidity, the percentage of H^+ ions in solution) between oceanic waters, the chemical processes are very similar in both polarity and quantity to those occurring in autotrophic living cells—those capable of directly producing organic matter like plants, without having to source it from the environment [102]. His ideas are compelling: they integrate all recent discoveries and even allow explaining how the structural and

E. Mieli et al., *The Living Galaxy*, https://doi.org/10.1007/978-3-031-67324-5_7

evolutionary differences between bacteria and archaea could have arisen through subsequent environmental specialization from a common ancestor—the famous LUCA (Last Universal Common Ancestor) of all current life on Earth [32]. However, while correctly criticizing other theories which must be accounted for, his hypothesis presents a significant gap. If the synthesis of biological molecules, as well as oligomerization (production of short molecule sequences) is possible in hydrothermal systems, the issue of nucleic acid polymers (fundamental molecules for transmitting genetic information and producing proteins) remains. In fact, under appropriate laboratory conditions mimicking alkaline hydrothermal vents, only suitably activated adenine manages to form short filaments of a few units. Non-activated adenine (AMP), instead, at most produces only dimers, while the other nucleotides do not polymerize [10]. In short, the molecules responsible for hereditary transmission, such as DNA, would have difficulty forming according to Lane's model.

For this reason, in this context, we will consider the hypothesis advanced by the Italian physicist Mario Ageno [2] in the first half of the 1990s. Although some of the processes hypothesized by Ageno still await experimental verification, they can theoretically allow for the production of triphosphate nucleotides—the active forms capable of reacting and forming nucleic acids (DNA, RNA, and hybrid segments). Moreover, triphosphate nucleotides, particularly ATP, are the main molecules that act as biological batteries, enabling biological chemical reactions.

As we will see, Ageno supposed that the first living beings were photosynthetic, because the only unlimited and omnipresent source of energy available on the face of the Earth, at least up to a certain marine depth, was provided by the Sun. The importance of electromagnetic radiation for present and past life is no longer in question, and photosynthesis carried out by green plants and cyanobacteria is the only process that allows the accumulation of free oxygen in the environment. It is therefore the only process that ultimately creates a barrier—the ozone layer—to the ionizing electromagnetic energy that reaches a planet's surface from space [57]. Consequently, in the absence of photosynthetic beings capable of releasing oxygen, the ozone layer does not develop in the atmosphere. Without an ozone layer, water is completely decomposed into its elementary compounds over time, eliminating one of the indispensable resources for life.

Therefore, without photosynthesis, the situation that occurred on Venus is determined, where the oceans have all been consumed by such a process. We thus arrive at the paradox that if living organisms capable of producing oxygen through photosynthesis do not form or are not themselves the first living beings, the life thus created fails to sustain itself and quickly vanishes.

For these reasons, we consider Ageno's hypothesis currently the most plausible to explain the necessary steps to evolve stable forms of life on a planet. Therefore, this will be the guiding thread that we will follow to evaluate the numerical variables to be used in the calculations of Maccone's algorithm, in the absence of new discoveries that cast doubt on its effectiveness.

8

The Theory of Mario Ageno

Ageno began by specifying his definition of a living being. For Ageno, a living being is an open chemical system (that is, a system in which chemical processes occur and which can exchange energy and materials), coherent (where the processes are ordered in space and time), and equipped with a program (possessing a "conductor", the DNA, which establishes what to do and when). Although the author stated that this is *One* possible definition for recognizing all terrestrial life forms rather than *The* definitive definition, based on his concept, it is easy to realize that the foundation of his idea of a living being is the biological cell, as for most modern specialists (see the previous discussion about Lane).

Building on these foundations, Ageno developed a hypothesis consisting of a series of stages, where the transition from one stage to the next was considered to occur with an almost certain probability (in a practically automatic manner). Only through such a process could the transition from non-living to living arise, given very precise initial conditions. The starting idea originated from the classic prebiotic broth and lipid sacs or lipid vescicles (bubble-like structures enclosed by molecules composed of fatty acids, hypothesized to transform into protocells) formed in a particular environment—the shallow substrate between **10** and **20 m** deep in a lagoon—ultimately leading to the formation of living cells.

Despite sharing the initial premise with other previous theories, Ageno introduced two important innovations that made his hypothesis much more credible than previous claims. Exploiting the properties of the components of lipid sacs, which can merge and split without mixing their contents with the external environment, Ageno emphasized that one did not have to track the

E. Mieli et al., *The Living Galaxy*, https://doi.org/10.1007/978-3-031-67324-5_8

fate of *a Single* sac. Instead, the system capable of evolving into a living being was constituted by the collective *Set of All* the lipid sacs present in the lagoon. If a particular reaction occurred in a certain sac, following random collisions between sacs, the product could transfer to another sac, inside which the next reaction could then take place. This greatly increased the probability of the entire process unfolding.

Ageno's second hypothesis consisted of the mechanism providing energy for the system to function. According to Ageno, the energy came from a simplified version of the photosynthesis of green plants: a lipid sac would trap chlorophyll pigments capable of becoming excited by photons and losing electrons along a simple *redox chain* (a series of processes describing the transfer of electrons from one chemical species to another), embedded in the membrane thickness and comprising a molecule with carbon-carbon double bonds. This would transfer protons H^+ inside the sacs, making the system acidic. The acidity would have favoured the maintenance of polyphosphates, including biological energy batteries, molecules such as ATP, and the formation of certain chains of organic molecules such as oligonucleotides (simple strands of DNA, RNA and/or nucleotides and other molecules such as amino acids) from the precursors present in the prebiotic broth. This would be the first step towards the formation of all the chemical processes that characterize living beings.

In response to objections that the photosynthesis of green plants, whose chlorophyll receives electrons directly from water, requires a complicated apparatus (including two photosynthetic pigments) and therefore cannot be considered primitive, Ageno replies that what is simpler is not necessarily older: a simple apparatus may be a localized adaptation of a system that was originally more complex. Recently, some studies are emphasizing the importance of photosynthesis for all phenomena related to life on our planet.

Let's now formalize mathematically the statement about collisions between lipid vesicles. We know that in a system of particles (our protocells) with cross-section **s** and density **n** per cubic meter, the mean free path (average distance traveled without collisions) is:

$$\lambda = 1/(\mathrm{n} \cdot \mathrm{s})$$

Therefore, defining **v** as the relative speed between particles, the average time between successive collisions will be:

$$t = \lambda / v = 1/(\mathrm{n} \cdot \mathrm{s} \cdot \mathrm{v})$$

t is interpretable as the *first reshuffling* time of the lipid vesicles. Knowing that:

$$s \sim 10^{-12}\,\mathrm{m}^2$$
$$v \sim 10^{-5}\,\mathrm{m/s},$$

it follows that:

$$t \sim 10^{17}/n$$

The latter value needs to be carefully controlled because concentrations less than $\mathbf{n} = \mathbf{10^8}$ particles/ $\mathbf{m^3}$ would produce reshuffling times greater than **30** years, which could be too long in certain variable and traumatic conditions hosting the protocells. However, it is also true that for a modest production of lipids distributed on the sea surface, for each $\mathbf{m^2}$ of lipids there could be about $\mathbf{10^{12}}$ lipid vesicles which, at a maximum depth of about ten meters (allowing solar energy penetration as we will see), would determine $\mathbf{n} = \mathbf{10^{11}}$. This is reassuring since in this case the reshuffling time would be $\mathbf{t} = \mathbf{10^6\,s}$, or about **10** days.

We will see later how marine meteorological agents, periodically interfering with the concentration of lipid pockets, often cause real temporal barriers (*microcatastrophes*) for reactions inside the pockets to be realized.

9

The Calculation of the Fourth Parameter f_l

We report the entire calculation process in Fig. 9.1, now describing the steps in detail:

Step 1: This initial step assigns frequencies starting from known or comparable data: each phase **j** represents a necessary process for life's development associated with a random variable $\mathbf{x}_j$ (fraction or frequency) to be estimated within the maximum and minimum values $\mathbf{a}_j$ and $\mathbf{b}_j$ over the observation time $\mathbf{\Delta T}_j$. The phase must be realized within a microcatastrophe time limit $\mathbf{\Delta T}_{0j}$, beyond which it consolidates. Ultimately, the four input values $\mathbf{a}_j$, $\mathbf{b}_j$, $\mathbf{\Delta T}_j$ and $\mathbf{\Delta T}_{0j}$ must be provided for each phase.

Step 2: In this step, the assigned frequencies are each remodulated within their time limit $\mathbf{\Delta T}_{0j}$ through the time covariant transformation:

$$X_j\left(m_j\right) = 1 - \left(1 - x_j\right)^{mj}$$

This transforms the frequency $\mathbf{x}_j$ of an event, related to $\mathbf{\Delta T}_j$, into the frequency $\mathbf{X}_j$ related to an interval $\mathbf{\Delta T}_{0j}$ which is $\mathbf{m}_j$ times larger, where $\mathbf{m}_j = \mathbf{\Delta T}_{0j}/\mathbf{\Delta T}_j$. If $\mathbf{x}_j$ is the single trial success probability, $\mathbf{X}_j$ is the consecutive $\mathbf{m}_j$ trials success probability. The same applies to minimum/maximum frequencies $\mathbf{a}_j$, $\mathbf{b}_j$ transformed into $\mathbf{A}_j$, $\mathbf{B}_j$. The sum of all $\mathbf{\Delta T}_{0j}$ gives the overall medium-term time $\mathbf{\Delta T}_0$ for the entire process:

E. Mieli et al., *The Living Galaxy*, https://doi.org/10.1007/978-3-031-67324-5_9

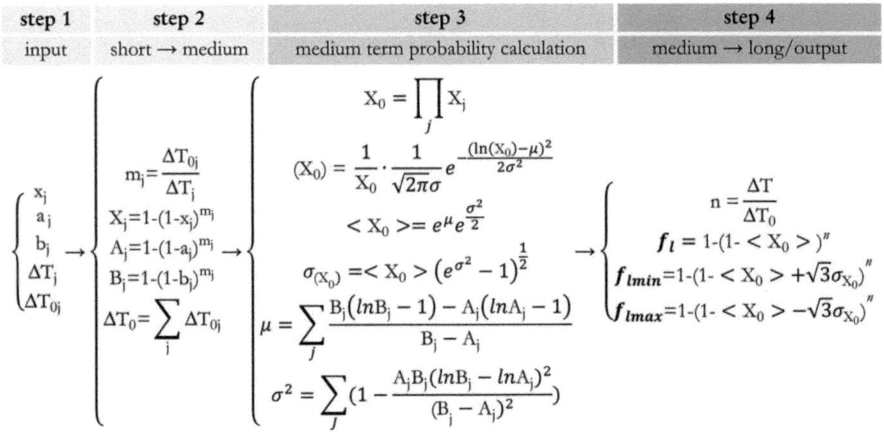

Fig. 9.1 4th Drake—from step **1** to step **4** the information on the individual processes in the short term give rise to the probability in the medium and long term **(IJA 14/06/2023 Mieli, Valli, Maccone)**

$$\Delta T_0 = \Sigma_j \Delta T_{0j}$$

Step 3: In the medium-term, phase frequencies $\mathbf{X_j}$ being temporally homogeneous, can be multiplied together, as per Drake's formula, to obtain the new medium-term variable $\mathbf{X_0}$ whose distribution is calculated with the Maccone lognormal:

$$\Phi(X_0) = \frac{1}{X_0} \cdot \frac{1}{\sqrt{2\pi}\sigma} e^{-\frac{(\ln(X_0)-\mu)^2}{2\sigma^2}}$$

where $\boldsymbol{\mu}$ and $\boldsymbol{\sigma}$ are the mean and standard deviation of the logarithm of $\mathbf{X_0}$, while $\mathbf{<X_0>}$ and $\boldsymbol{\sigma}(\mathbf{X_0})$ are the mean and standard deviation of $\mathbf{X_0}$.

Step 4: Similarly to step 2, the obtained frequency $\mathbf{<X_0>}$ is remodulated in the macrocatastrophe time limit $\mathbf{\Delta T}$ through the same *covariant* algorithm:

$$f_l(n) = 1 - (1 - X_0)^n$$

where n = $\Delta T / \Delta T_0$

The two deviations from the mean $\mathbf{X_0} \pm \sqrt{3}\,\boldsymbol{\sigma}(\mathbf{X_0})$ give rise to $\mathbf{f_{l\ min}}$ and $\mathbf{f_{l\ max}}$, derived from Maccone's lognormal formula (Appendix **D**).

In conclusion, providing the four input values $\mathbf{a_j}$, $\mathbf{b_j}$, $\Delta\mathbf{T}_j$ and $\Delta\mathbf{T}_{0j}$ for each phase, the final probability $\mathbf{f_l}$, between the minimum and maximum values, $\mathbf{f_{l\ min}}$ and $\mathbf{f_{l\ max}}$ over time $\Delta\mathbf{T}$ is obtained.

10

The Transition from Non-living to Living, Phase by Phase

The phenomenon of the onset of life linked to the parameter $\mathbf{f_l}$, has been divided into **ten** phases, identified below:

1. The abiotic synthesis of biological molecules
2. The concentration of the primordial broth
3. The formation of lipid sacs
4. The inclusion of chlorophyll in lipid membranes
5. The "proton photopump"
6. The formation of nucleic acid filaments
7. The catalytic role of RNA
8. Determination of roles
9. Formation of the cell membrane
10. Emergence of the genetic code

10.1 The Starting Point; the Habitable Planet

The basic conditions for developing the transition from non-living to living on an Earth-like planet, as outlined in the first three Drake parameters, are summarized below:

E. Mieli et al., *The Living Galaxy*, https://doi.org/10.1007/978-3-031-67324-5_10

(a) The planet must be rocky and not gaseous, like the inner solar system planets Mercury, Venus, Earth, and Mars.
(b) It must be at an appropriate distance from its star to allow liquid water to exist on the planet's surface. Water is indispensable for life as we know it, serving as the preferred solvent for metabolic processes.
(c) The planet's gravity must be sufficient to retain an atmosphere that acts as a protective shield against cosmic rays and ultraviolet radiation. The conditions described above are continuously monitored and catalogued on thousands of exoplanets that have been detected since the second half of the 1990s.
(d) The atmosphere should not be oxidizing like Earth's current atmosphere, but rather devoid of free oxygen (**O_2**). The stable, significant presence of O_2 results from photosynthetic processes of living organisms. While water can be split into hydrogen and oxygen by electromagnetic radiation, producing trace amounts of **O_2**, any oxygen produced tends to quickly oxidize with molecules in the environment. Therefore, a celestial body without living beings is believed to have an atmosphere lacking free oxygen or containing only trace amounts.
(e) Finally, the planet must have geothermal activity, with internal heat moving towards and being released at the surface through phenomena like volcanism and hydrothermal vents [103] (Fig. 10.1).

Fig. 10.1 Eo-Archean environment (about **4–3.6 Gy**) Powered by OpenAI

With this background established, we can now look in more detail at the various phases of the process transitioning from non-living to living conditions.

10.2 The First Phase; the Abiological Synthesis of Biological Molecules

To initiate the process, the abiotic synthesis of biological molecules and/or their precursors is essential. It is now known that under appropriate conditions (presence of energy and suitable chemical precursors) and in the presence of a reducing atmosphere, simple organic molecules like lipids, amino acids, nucleotides (Fig. 10.2), hydrocarbons and others can be produced [79, 91].

While the early Earth did not have a completely reducing atmosphere, it was devoid of free oxygen. On a planet with the listed characteristics, organic compounds can derive from meteoritic contributions of molecules produced in space (during the Archean eon between **4** and **2.5 billion** years ago, hundreds of thousands of meteorites fell on Earth) and from synthesis in hydrothermal vents [77]. It is undeniable that such a planet can contain organic molecules.

To estimate the minimum and maximum frequency of biological molecule formation, the slower and less probable of the two described phenomena—meteoritic impacts—is taken as reference. During the Hadean and Archean, the meteorite impact frequency on Earth was significant. Recent work by Takeuchi and collaborators [120] quantifies around $\mathbf{10^{20}}$ **kg** of meteorites arriving in that period—a huge value (the terrestrial mass is about $\mathbf{6 \cdot 10^{24}}$ **kg**) amply justifying the presence of the molecules in question.

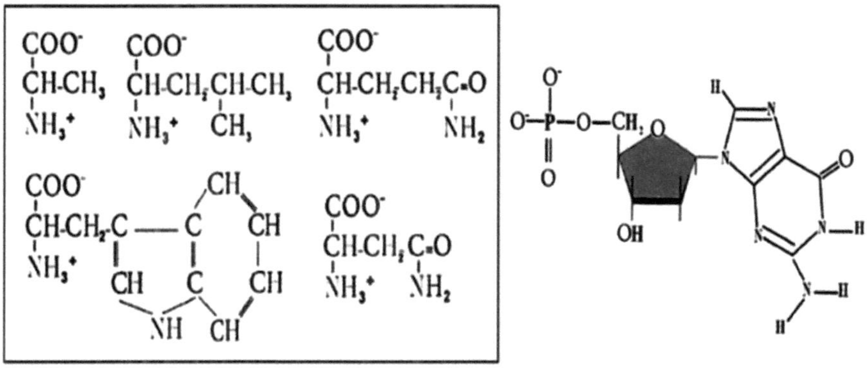

Fig. 10.2 Some of the biological molecules necessary for life: in the box on the left, **5** of the **20** fundamental amino acids (at the top, from left to right, Alanine, Leucine, Glutamine; at the bottom, from left to right, Tryptophan, Asparagine); on the left, the monophosphate nucleotide obtained with the Guanine **(IJA 14/06/2023 Mieli, Valli, Maccone)**

We hypothesize that the distribution of meteoritic biological molecules affected between 90% (value a_1) and 100% (value b_1) of the planet over an observation time ΔT_1 of about 1,000,000 years. Furthermore, we will keep the maximum ΔT_{01} (*microcatastrophe* time of phase 1) equal to 1,000,000 years, considering the existence conditions of hydrothermal vents likely have a limit value of that order of magnitude.

Phase 1

a_1	b_1	ΔT_1	ΔT_{01}
0.9	1.0	1,000,000	1,000,000

10.3 The Second Phase; the Concentration of the Primordial Broth

Ageno situates the evolution of his system in a particular environment: the bottom of a sheltered lagoon, at a depth between **10 and 20 m**. Beyond the details, what is important to underline is that the concentration of organic compounds must be consistent with a depth at which harmful electromagnetic rays (those too energetic, capable, that is, of destroying biological molecules and altering chemical reactions) are absorbed by the liquid, but such that that part of the spectrum used for photosynthesis is still accessible. The depth, therefore, depends on the particular composition of the atmosphere and the intensity of the light that reaches the surface of the planet (Fig. 10.3).

How realistic is this picture? In a completely sterile environment (total absence of living forms) it is difficult to imagine phenomena capable of altering and destroying molecules, beyond ultraviolet (UV) rays and other energetic electromagnetic radiations which, however, are absorbed by water molecules beyond a certain depth. Moreover, if we add the proximity to an alkaline hydrothermal source, we can be sure of having secured a regular supply of organic material.

Studies related to hydrothermal sources show that, if the majority of them are located between **2000** and **3000 m** below sea level, they are actually also abundant in subaerial environments at various other depths. And this was likely the case in the past as well. We can therefore consider that during the Archean, alkaline hydrothermal sources were present and widespread on our planet.

To estimate the probability of the presence of biological molecules in alkaline hydrothermal sources we will take a minimum frequency equal to 5%

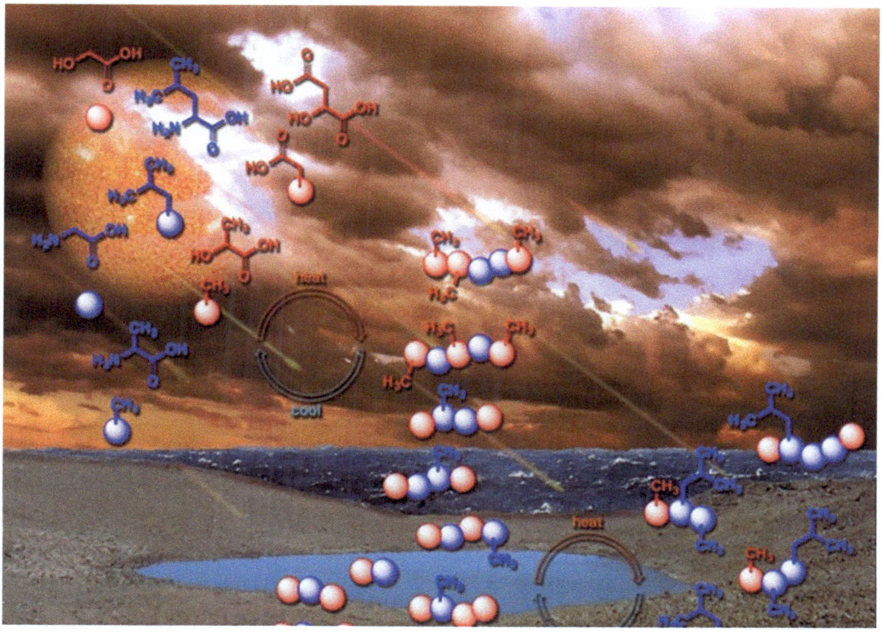

Fig. 10.3 Sheltered lagoons that favor the concentration of the primordial broth **(Ram Krishnamurthy—Center for Chemical Evolution—Scripps Research Institute)**

and a maximum equal to 15%. These values are suggested by the distribution of lagoons with such depth characteristics over a release time of about 100 years. The maximum lifespan of lagoons in such optimal conditions is estimated at 10,000 years.

Phase 2

a_2	b_2	ΔT_2	ΔT_{02}
0.05	0.15	100	10,000

10.4 The Third Phase; the Formation of Lipidic Vesicles

Among the organic molecules present in the defined environment, some particular fatty acids stand out: lipids with a polar head and a hydrophobic tail.

These compounds are able to form double-layered molecular layers on the surface (Figs. 10.4, top and 10.5) and, when immersed in water, double-walled pouches (Figs. 10.4, sides and 10.5) in which the polar ends of the molecules are in contact with the water while the hydrophobic ends (which

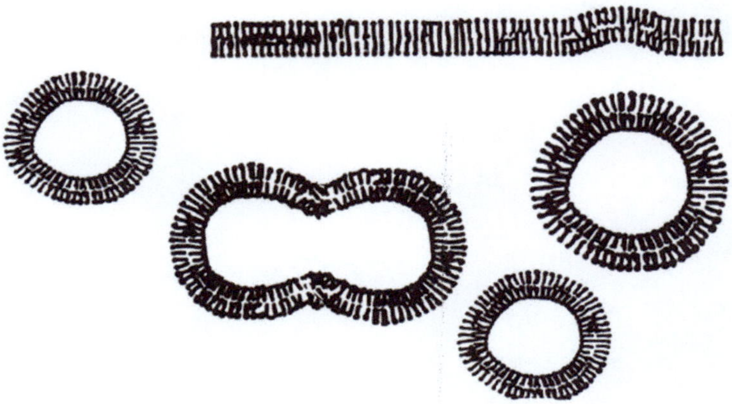

Fig. 10.4 Above, double-thickness surface layer; on the sides double-walled lipid vesicles; in the center, fusion of two lipid vesicles **(IJA 14/06/2023 Mieli, Valli, Maccone)**

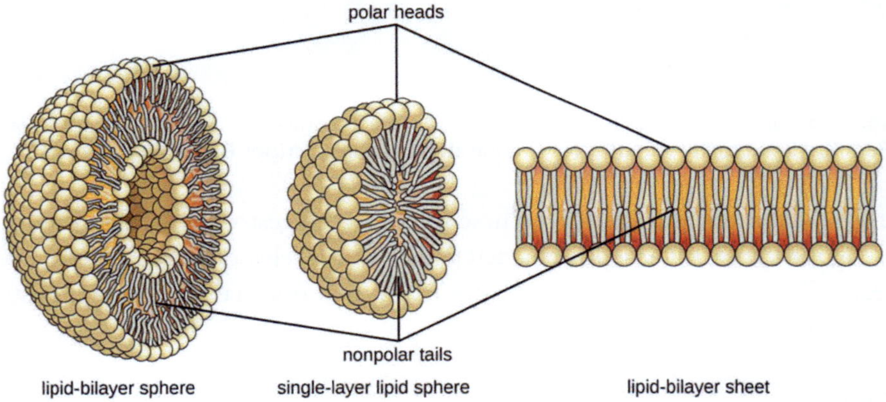

Fig. 10.5 Section of a lipid vesicle with a double-thickness surface **(CNX OpenStax)**

do not like this liquid) face each other in the center of the wall thickness. These molecules are among the main components of organic membranes, particularly those of cells that make up all living beings.

The double-walled lipid pouches thus formed are able to fuse (Fig. 10.4, center), upon collision, and to split in two if the volume of one is excessive, without mixing their content with the external environment. In this way, it is possible to divide our system into two distinct spaces, the **external** one of the lagoon and the **internal** one, which, as we have seen, statistically includes the space in *All* the present pouches.

The pouches formed in the lagoon can incorporate some of the abiotically synthesized organic material. If a certain compound is present inside a pouch, due to the processes described, it can at a certain moment come into contact with another, initially produced in a different pouch, and react with it. In fact, due to the fusions and separations, the internal environment is unique and any molecule can sooner or later interact with the others.

The formation and dynamics of the vesicles depend on the lipid concentration and physical characteristics of the liquid (water) in which they are immersed. Since lipids can be easily synthesized in hydrothermal sources, they must have been relatively abundant. At this point, given their behavior, the formation of lipid bilayers and vesicles was relatively probable and rapid: we set the frequency between 0.9 and 1 in the equally rapid time of 0.1 years. The time limit, or environmental crises capable of destroying the system, were extreme climatic events disturbing the lagoon tranquility, estimated at about 10 years.

Phase 3

a_3	b_3	ΔT_3	ΔT_{03}
0.9	1.0	0.1	10

10.5 The Fourth Phase; the Inclusion of Chlorophyll in Lipid Membranes

The prebiotic broth did not only contain amino acids, nucleotides or lipids. Abiotic synthesis also allowed for obtaining other types of molecules, more or less simple. For example, laboratory experiments like Miller's have demonstrated the abiotic synthesis of chlorophyll, a pigment that allows green plants and cyanobacteria to perform photosynthesis by exploiting electrons subtracted from water molecules [23]. Although complex, this molecule could be very ancient: fossils of organisms capable of photosynthesis like green plants have been found in **3.5 billion** year old Australian sedimentary layers [97].

Chlorophyll is a pigment (molecule) that can get excited if stimulated by photons of a suitable wavelength. The excitation causes the loss of one or more electrons that begin to travel a *redox chain* (a series of electron transfer processes between chemical species), formed by several molecules contained in the photosynthesizing membrane thickness. Among these, the quinones stand out, particular molecules that, by reducing (accepting electrons), must bind to an equal number of protons (**H^+**). One peculiarity of these molecules is that they possess alternating double bonds responsible for this characteristic.

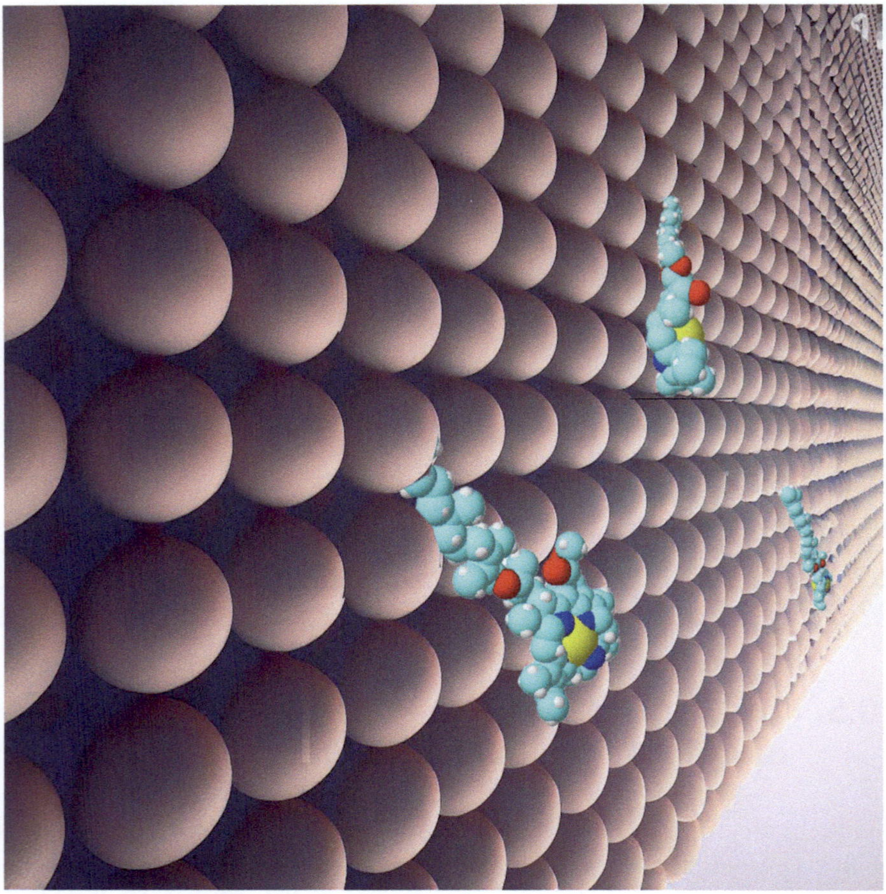

Fig. 10.6 The trapping of chlorophyll and other components of the redox chain in the membrane Powered by OpenAI

The simplest alternating double bond molecule, therefore having quinone properties, is much simpler than many other abiogenic synthesis compounds. Therefore, examples of this substance were likely present in the prebiotic broth. Since certain quinones are liposoluble, capable of mixing with fats especially in biological processes, some could have been "trapped" in the membrane thickness of lipid sacs formed in the prebiotic broth, preserving their essential redox chain property: the reversible reduction accompanied by reversible proton acceptance. If the temperature was not too low, such molecules could move within the lipid membrane like a two-dimensional liquid, like the quinones of photosynthetic systems.

The trapping of chlorophyll (Fig. 10.6) and other redox chain components in the membrane would have allowed the passage of an electron flow powered by sunlight [3]. It is important to emphasize that for this device's operation, the entire redox chain formation is not necessary, but only the presence of elements allowing electron passage through the quinone to carry out proton transfer. Chlorophyll and necessary redox chain molecules were certainly present in the primordial broth. Being liposoluble, they could penetrate the double lipid layer and remain trapped.

Assuming not too high a chlorophyll concentration, we can assign a frequency between 10 and 20% of pigment inclusion for an observation time of about 1 year. The phase can be compromised if the lipid sac system is diluted; as in the previous case, we will set the time limit at 10 years.

Phase 4

a_4	b_4	ΔT_4	ΔT_{04}
0.1	0.2	1	10

10.6 The Fifth Phase; the "Proton Pump for Photosynthesis"

Once present in the thickness of the lipid membrane, the set of molecules indicated above allows Ageno's "proton photopump" to come into action. What is it? A mechanism to subtract protons (**H**$^+$ ions) from the external environment and transfer them inside the sac system.

The juxtaposition of a pigment—in our case, chlorophyll, because it allows obtaining electrons from water molecules—with an appropriate double bond molecule, will allow, when the pigment receives the appropriate photons, to pass the excited electrons to the double bond molecule. The resulting polarity will serve to keep the two molecules close together, inside the lipid double layer. And so on, for all molecules part of the redox chain.

The reduction of the double bond molecule is accompanied by acquisition of protons from the *External* environment that, upon subsequent electron release towards another substance, are given to the *Internal* environment instead.

Therefore, there is a chance that a system capable of transferring protons into the internal environment, making it acidic, may be generated. Naturally, there is also a chance the opposite may occur.

However, in the first case, the system can evolve in the direction indicated by Ageno; in the second, it stops. Given the prebiotic broth concentration, it is likely more pigments and double bond molecules are trapped in the lipid membranes. Those concentrating protons inside (Figs. 10.7 and 10.8) will evolve towards the next phases, the others will have no future.

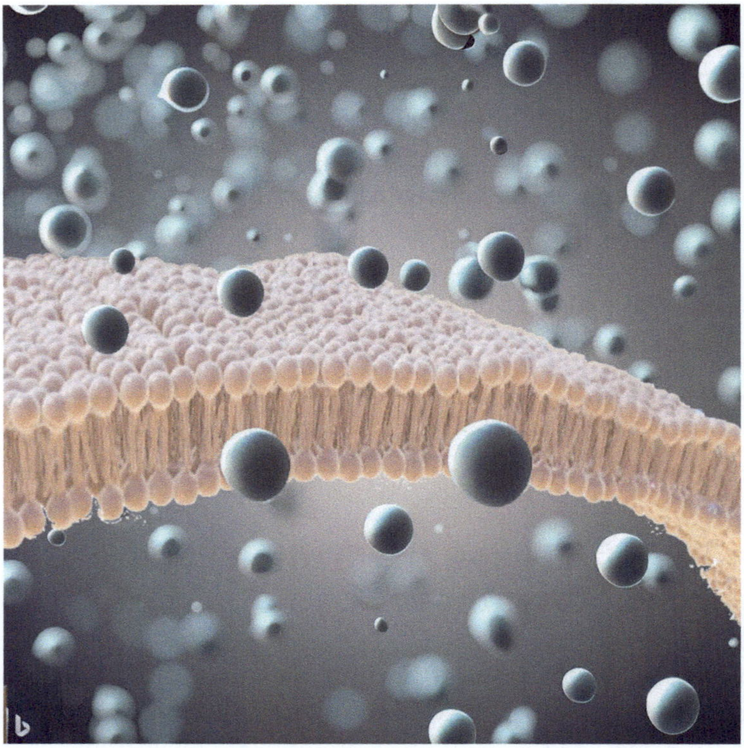

Fig. 10.7 Asymmetric passage of protons through the membrane Powered by OpenAI

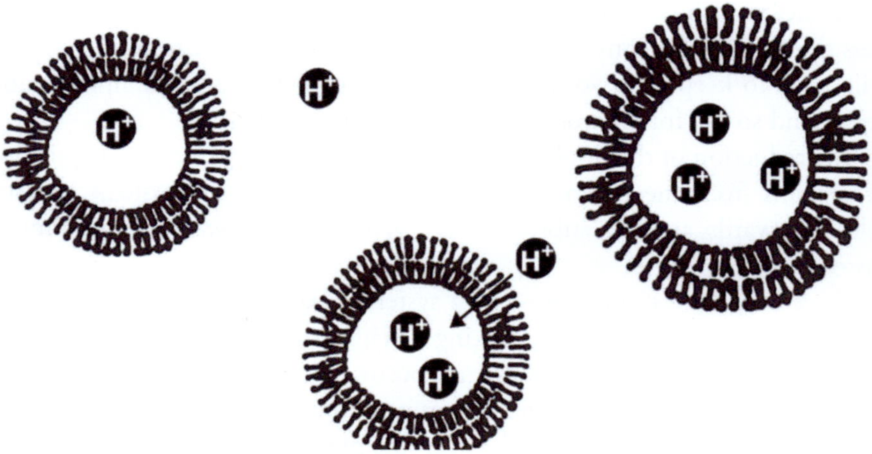

Fig. 10.8 The concentration of $\mathbf{H^+}$ ions within the system of lipid sacs increases the pH of this environment: the arrow indicates the entry of a $\mathbf{H^+}$ ion thanks to the "proton photopump" **(IJA 14/06/2023 Mieli, Valli, Maccone)**

Also bear in mind the primordial ocean was presumably more acidic than current, implying a higher proton concentration that may have contributed to the pump working in the right direction.

Let's translate this into a formal statistical phenomenon: chlorophyll, initially, with **50%** probability **p**, acidifies either inside or outside the lipid sac; however, this probability will have a binomial standard deviation of **25%**, or:

$$N \cdot p \cdot (1-p) = N \cdot 1/2 \cdot 1/2 = N/4$$

where **N** is the photon absorptions in the observation time; it will be this deviation from the average that makes part of the more acidic internal lipid sacs able to continue the process.

Naturally, lipid sac fusions and separations can reshuffle contents, but it is undeniable Ageno's photopump would have been able to "acidify" at least part of the internal system stably. Let's assume this fraction is between 10 and 20% in an observation time of a few days, or 0.01 years and a 10 year limit time as in the two previous phases.

Phase 5

a_5	b_5	ΔT_5	ΔT_{05}
0.1	0.2	0.01	10

10.7 The Sixth Phase; the Formation of Nucleic Acid Filaments

The acidification of the internal sac environment, or at least a part of it, allows solving one of the most important problems related to the transition from non-living to living: the precipitation of insoluble phosphorus apatites. Phosphorus is essential to life, intervening in nucleic acid (DNA, RNA) formation and biological energy batteries (the most important being ATP, the triphosphate form of the adenosine nucleotide, an RNA component). It is precisely the triphosphate nucleoside forms that are activated and can therefore react to form nucleic acid strands. If the trifosfato forms are not preserved, nucleic acid construction is lost. This is the crucial step that made us prefer Ageno's hypothesis to Lane's.

The acidification of the internal environment allows, therefore, the conservation of polyphosphates and organic phosphates produced in an abiotic way. In particular, it favours two essential elements for evolution towards the living:

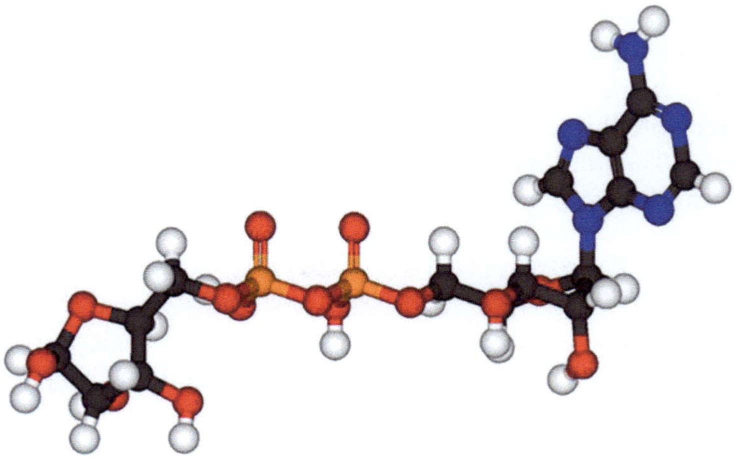

Fig. 10.9 Molecule of ATP

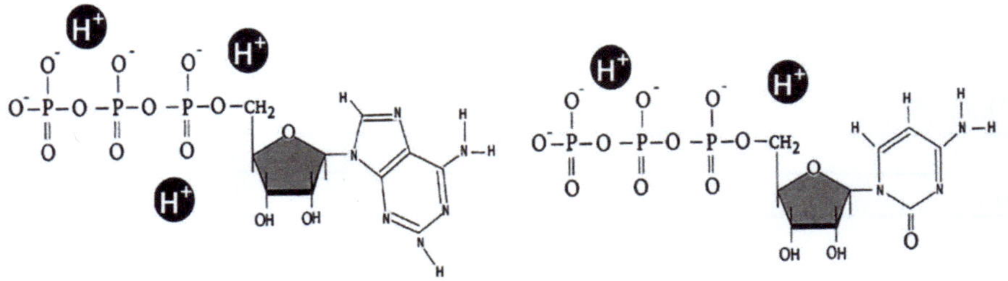

Fig. 10.10 Triphosphate nucleotides formed in the environment made acidic by $\mathbf{H^+}$ ions (ATP, with the adenine base, on the left; CTP, with the cytosine base, on the right) **(IJA 14/06/2023 Mieli, Valli, Maccone)**

1. The formation, starting from prebiotic broth basic components sequestered inside the lipid sacs, of DNA, RNA or nucleotide/amino acid hybrid strands
2. The preservation of ATP formed in an abiotic environment (Figs. 10.9 and 10.10, left), the most important biological "battery" capable of allowing endoenergetic chemical reaction evolution

Naturally, the lipid sac dynamics, with fusions and divisions, allowed new material processed in the more acidic compartments to be available throughout the internal environment.

Given the complexity of this phase, we will assign it a low 2–3% probability over 0.01 years (3–4 days) and a 1 year time limit beyond which seasonal variations could obstruct phase completion.

Phase 6

a_6	b_6	ΔT_6	ΔT_{06}
0.02	0.03	0.01	1

This also allows accounting for the fact that current ATP production involves molecules and processes not yet present at the time. What is really important is showing the possibility of stably maintaining triphosphate forms to allow nucleic acid molecule formation.

10.8 The Seventh Phase; the Catalytic Role of RNA

The production of strands of nucleic acids (DNA, RNA and mixed chains of amino acids and nucleotide acids) and of triphosphate nucleotides in the "photopump" system allows, gradually, the further evolution of the system.

Starting from the strands of nucleic acids, replicative processes can begin dependent on the particular combinations present. By trial and error, certain strands have become more abundant than others. In parallel, the enzymatic activity, initially performed by more or less isolated metal ions, becomes the prerogative of more complex molecules, including organic components.

Is this, perhaps, the famous *RNA world* [81], in which this molecule performed both enzymatic tasks and those related to inheritance (Fig. 10.11)? We do not want to go into details. We only remember that RNA, not being able to form the double helix, is ill-suited to preserve genetic information, unlike DNA. Moreover, there is no reason to believe that the two types of nucleic acids could not have evolved in parallel or from a common ancestor. Since in the current living world, the hypothetical RNA world has left no traces, we can reasonably consider DNA and RNA as two types of competing filaments, but which quickly ended up assuming different biological roles where DNA, with the double helix made up of deoxyribonucleotides, has become the molecule with a hereditary function.

On the other hand, the importance of RNA in this phase of the process of the transition from non-living to living should not be underestimated. Research has highlighted an increasing number of RNA sequences capable of enzymatic activity. All this suggests that, at least initially, it was RNA that performed the main enzymatic actions [17]. Let's also remember that in the more acidic environments of the internal compartment of the sacs, triphosphate nucleotides could form capable of providing the energy for the

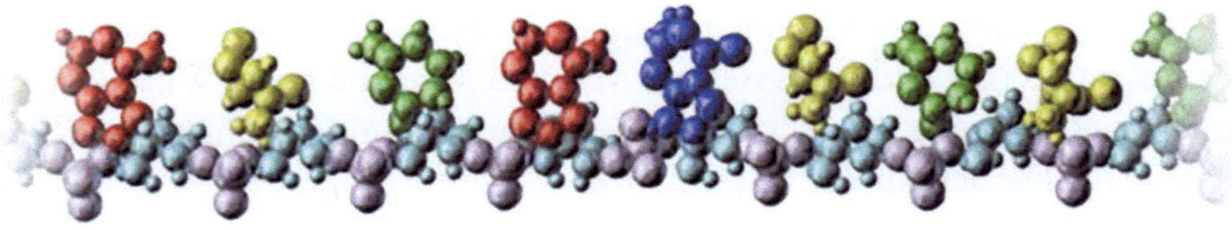

Fig. 10.11 Fragment of RNA strand

synthesis of new molecules and for allowing the elongation of nucleic acid filaments, the most active from the catalytic point of view.

It is now practically certain that RNA filaments can function as biological catalysts: the longer the filaments, the easier it is to find a reaction that is catalyzed by these even if, naturally, being less specific, they are less effective than proteins. If the concentration is high, the times of the chemical reactions are quite short (hours), less short if the concentration is low (days-months). We will put ourselves in the second case by placing the fraction of suitable RNA molecules between 90 and 100% in a time of 0.1 years (about a month) with a limit of about 1 year, as in the previous phase.

Phase 7

a_7	b_7	ΔT_7	ΔT_{07}
0.9	1.0	0.1	1.0

10.9 The Eighth Phase; Determination of Roles

Our system, now equipped with catalytic molecules and energy batteries, was able to start producing new biological molecules from a suitable carbon source. Having evoked photosynthesis as the main engine of our process, it is natural to consider carbon dioxide (**CO_2**) as such a source.

On the other hand, this liposoluble molecule could overcome the double lipid membrane and diffuse into the internal sac environment. Moreover, at the time, the **CO_2** rate was much higher than the current one. Carbon dioxide was well suited as a carbon source for the processes evoked previously and is considered to occupy such a role also in Lane's theory.

Little by little, through trial and error, the roles of the different organic molecules present began to differentiate. DNA, thanks to its double helix properties (Fig. 10.12), established itself as the molecule responsible for inheritance and control of the entire system. Proteins, thanks to their particular three-dimensional structure specific for each (Fig. 10.13), have proven to

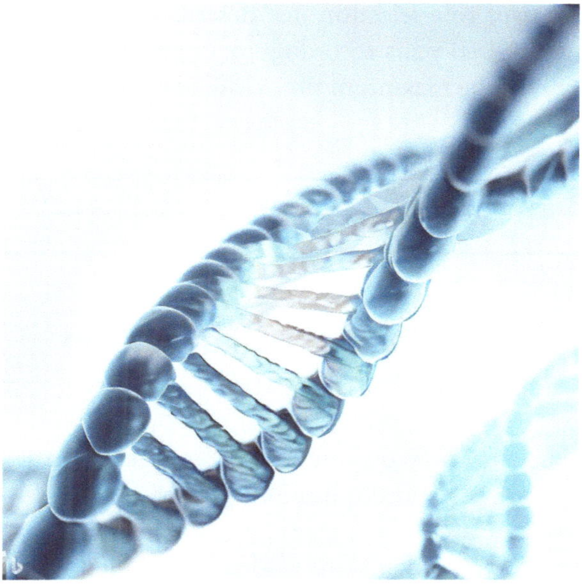

Fig. 10.12 Representation of a double helix DNA strand Powered by OpenAI

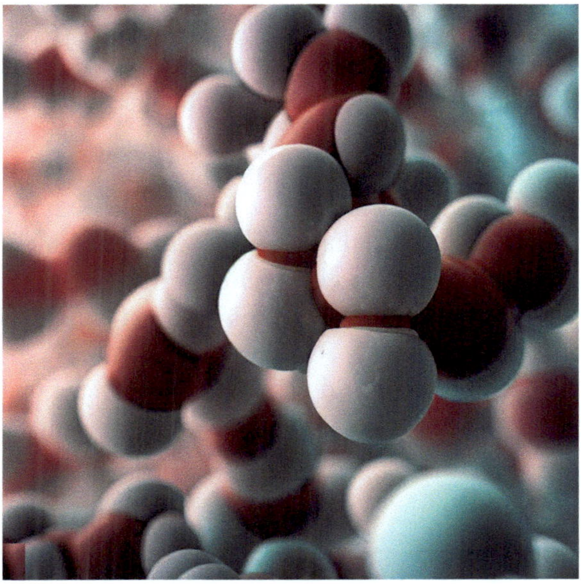

Fig. 10.13 Three-dimensional representation of a protein Powered by OpenAI

be more functional, distinguishing themselves for their extreme specificity in performing enzymatic activity. Finally, RNA specialized as an intermediary between DNA and protein synthesis, becoming indispensable for this activity (a significant part of ribosomes is composed of RNA).

The phenomena of role selection just described could have been realized gradually over a hundred years according to a maximum and minimum fraction of 20 and 30%. The maximum time coincides with the observation time.

Phase 8

a_8	b_8	ΔT_8	ΔT_{08}
0.2	0.3	100	100

10.10 The Ninth Phase; the Formation of the Cell Membrane

Following the last step, new organic molecules spread in the system. Those concerning the synthesis and duplication of particular DNA chains will be favored and propagate.

However, new protein chains are gradually included in the double lipid thickness. Some maintain its stability and coherence; others allow production of triphosphate nucleotides (the bases for producing DNA, RNA as well as biological batteries) from their precursors. The pouch in which all this happens will tend not to merge with others to exchange content and will acquire characteristics allowing it to house the system to produce triphosphate nucleotides from precursors (to easily produce the biological batteries, as in current bacteria).

Little by little, the various metabolic processes begin accumulating in single pouches that end up acting independently. Finally, at least one of these, of the appropriate size, will include all the processes and achievements listed in the previous stages, which can occur in a single protected microenvironment (Figs. 10.14 and 10.15). It will then be this last one which will continue its evolution towards the living.

We can assign low minimum and maximum frequencies of 1 and 2% over an observation time of a few days (protein chain inclusion time in the membrane), or 0.01 years, and a 1 year limit time (seasonal cycle).

Phase 9

a_9	b_9	ΔT_9	ΔT_{09}
0.01	0.02	0.01	1

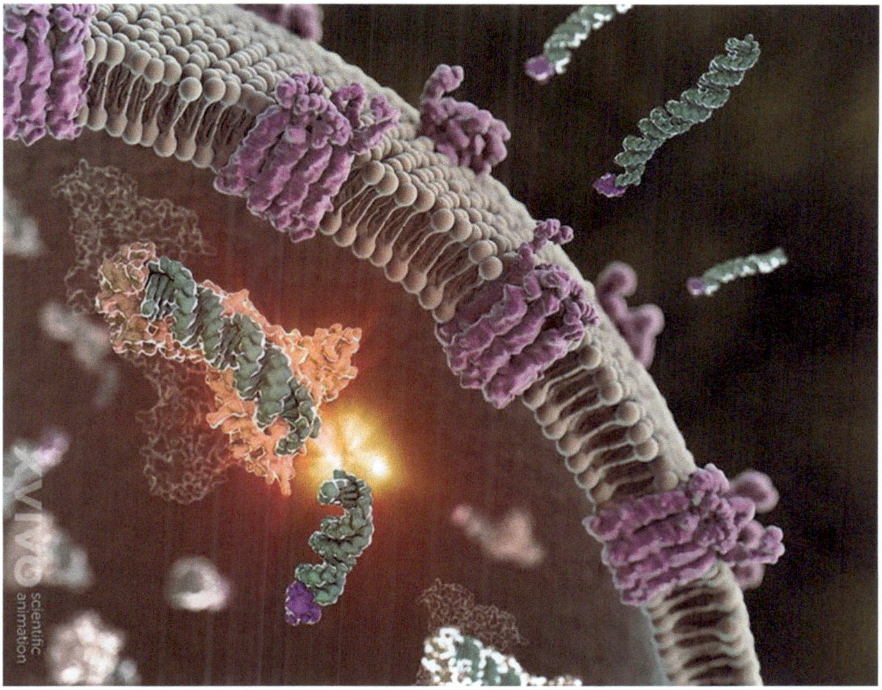

Fig. 10.14 Representation of a cell membrane (**Nastech**)

Fig. 10.15 The metabolic processes of the previous stages now all take place in a single sac, whose surface no longer merges with the others (**IJA 14/06/2023 Mieli, Valli, Maccone**)

10.11 The Tenth Phase; the Emergence of the Genetic Code

Once the processes described in the previous phases have been established, there is still a fundamental step before obtaining a living being: the emergence of the genetic code, that is, the appropriate combinations of three nucleotides (Fig. 10.16) capable of specifically indicating the amino acids to be added to the protein chain in the previously seen protocell.

It is in this way that the main function of the "program" (which dictates what to do and when) constituted by the DNA can exercise its operation. And it is also in this way that coherence can be established in the chemical processes of the system.

How is this step achieved? Through evolutionary trial and error [62], the various nucleic strands that favor the amino acid chains that contribute to the maintenance and generation of other copies of the same strands will be favored and perpetuated. Thus, gradually, in the protocell, one or some DNA chains will be able to produce all the molecules necessary for their duplication: these chains spread and every change capable of improving or accelerating the process is selected by evolution.

With the completion of this final stage, we have reached a system that is now able to satisfy both Ageno's conditions and the six needs indicated by Lane, within a single compartment, now modified compared to the starting sacs. We have finally obtained our living being.

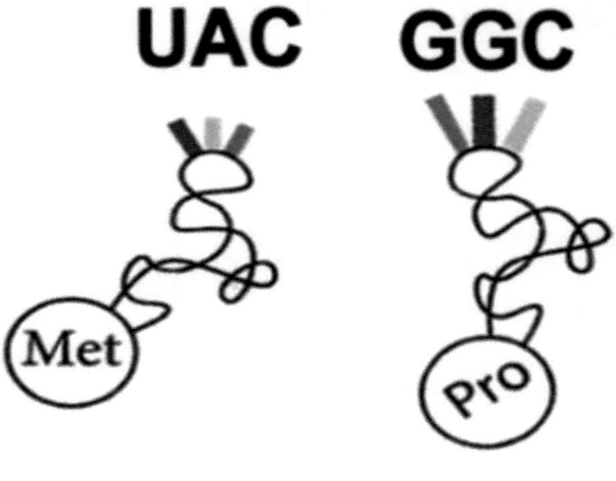

Fig. 10.16 Examples of "translation" of the genetic code: on the left, the triplet "Uracil-Adenine-Cytosine" (UAC) codes for the amino acid Methionine (Met); on the right, the triplet "Guanine-Guanine-Cytosine" (GGC) codes for the amino acid Proline (Pro) **(IJA 14/06/2023 Mieli, Valli, Maccone)**

In this last case, as far as estimating frequencies is concerned, we face two obstacles: the formation of nucleotides (from a phosphate group, a pentose sugar, and a nitrogenous base) and their combination into strands. While the second process seems very likely, the first is not as much. Focusing on this, as in the previous case, we can assign minimum and maximum low frequencies of 1 and 2%, this time, however, over an observation time of 5 years and a limit time of 20.

Phase 10

a_{10}	b_{10}	ΔT_{10}	ΔT_{010}
0.01	0.02	5	20

Before concluding, we point out that this phase can be reversed with the previous one. In fact, the emergence of the genetic code, the final step toward the constitution of the living being, may have occurred when the system was still widespread within all the pockets, or more than one of them.

What is the correct order? It is difficult to say with precision. By confining the last step to a single "pocket" and leaving it the task of making the transition from non-living to living, we ensure that all descendants will have the same genetic code, as is the case with all current organisms on Earth. We do not know if in the past there were multiple different genetic codes nor what their impact might have been on the evolution and interaction of living beings. We will return to this problem later, before addressing the further evolution of the system.

We note, however, that the pure inversion of two or more phases, without any modification of the different values attributed in the different stages, does not change the final value of the probability obtained from Maccone's algorithm, precisely because of the inherent mathematical properties of the algorithm itself.

10.12 Evaluation of Probabilities at Each Stage

We have thus obtained the **40** input values of step 1 reported in Table 10.1.

$\mathbf{\Delta T_0}$, as already written, is the sum of the $\mathbf{\Delta T_{0j}}$ and represents the average period; while the long-term limit $\mathbf{\Delta T}$ is set by us at about **100,000,000** years.

In conclusion, at the end of our journey, we have found, through the log-normal $\mathbf{\Phi}$, a probability of realizing a full cycle of the transition from non-living to living, in the average period $\mathbf{\Delta T_0 \sim 1{,}000{,}000}$ years, equal to about **0.7%,** (Fig. 10.17).

Table 10.1 4th Drake: the **40** values of the frequencies $\mathbf{a_j}$ and $\mathbf{b_j}$ minimum and maximum, of the observation time $\mathbf{\Delta T_j}$ and of the microcatastrophe time $\mathbf{\Delta T_{0j}}$, for each phase described in the previous paragraph

Phase	Description	a_j	b_j	ΔT_j	ΔT_{0j}
1	The abiogenic synthesis of biological molecules	0.90	1.00	1,000,000	1,000,000
2	The concentration of the primordial broth	0.05	0.15	100	10,000
3	The formation of lipid vesicles	0.90	1.00	0.10	10
4	The inclusion of chlorophyll in lipid membranes	0.10	0.20	1	10
5	The "proton photopump"	0.10	0.20	0.01	10
6	The formation of nucleic acid filaments	0.02	0.03	0.01	1
7	The catalytic role of RNA	0.90	1.00	0.10	1
8	Determination of roles	0.20	0.30	100	100
9	Formation of the cell membrane	0.01	0.02	0.01	1
10	The emergence of the genetic code	0.01	0.02	5	20

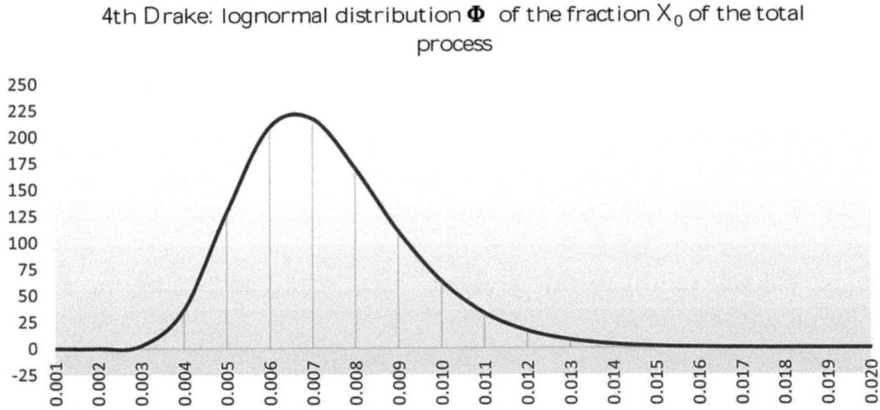

Fig. 10.17 4th Drake: the lognormal distribution Φ of the process in the medium period ΔT_0: the average value is **7.31** · **10**⁻**3**, the standard deviation is **1.95** · **10**⁻**3 (IJA 14/06/2023 Mieli, Valli, Maccone)**

This value, applying the rule of transformation of probabilities described for the *phases* (*covariant with time*), that is:

$$p_A = 1 - (1 - p_{A0})^n$$

translates, in the long term **ΔT~100,000,000** years, into an average probability of the onset of life on Earth equal to:

$$f_l = 52\%$$

between the two minimum and maximum values

$$f_{lmin} = 32\% \text{and} f_{lmax} = 66\%$$

Drake 4

$f_{l\,min}$	$f_{l\,max}$
$3.2 \cdot 10^{-1}$	**$6.6 \cdot 10^{-1}$**

11

Considerations on the Fourth Parameter

Although our work can only be considered preliminary—and certainly contains elements and conditions that will need to be reviewed in light of future scientific discoveries—we can already recognize some interesting results.

Firstly, we found a value for the probability of life arising within **100 million** years that, contrary to what one might expect, is on the order of **0.5**, or rather decidedly high (incidentally equal to the value inserted by Maccone in 2008 in his statistical Drake equation). Therefore, we do not need to invoke an *anthropic principle* to justify our presence as observers of the phenomenon of life on Earth (a reasoning along the lines of: life in the universe is very rare, but since I exist and represent life, the probability is nonetheless non-zero). Instead, we can fall within the principle of mediocrity, which asserts that the Earth, and we with it, is not a privileged point, and we are not special *even as observers*. This conclusion was far from taken for granted, even though recent paleontological observations already hinted that life on our planet formed as soon as it had the opportunity, a few tens of millions of years after the planet stabilized [19, 84].

Moreover, Maccone's method of dividing the problem into more individually approachable sub-problems from a mathematical-statistical point of view makes the phenomenon of the emergence of life less obscure or, at least, partially manageable. It is obvious that the frequency values attributed to the phases and the phases themselves can be improved and redefined (we hope that in the future they will be), but what matters is that the algorithm gives a response consistent with the input data. It would be interesting to explore the probability obtained using a different model—for example, that of Lane or

E. Mieli et al., *The Living Galaxy*, https://doi.org/10.1007/978-3-031-67324-5_11

others—for the transition from non-living to living, but this goes beyond the scope of our book.

Finally, we want to point out that there is no preclusion to using this approach not only to planets entirely similar to Earth but also to situations that deviate slightly without precluding the possibility of the emergence of life. We are thinking of some nearby situations like Saturn's sixth moon, Enceladus, which could present conditions favorable to life under its icy crust. But we are also thinking of more distant situations like terrestrial-type exoplanets located in the habitable zone of their star, like the Trappist-1 system at **40 light-years** from us; in such situations, we could have planets with synchronous rotations (rotation equal to revolution) or super Earths with masses greater than **10** times that of Earth.

To conclude, it is necessary to point out, once again, that for now we have dealt with the emergence of life in its most basic form: unicellular and prokaryotic. It is in this form that, with the appropriate initial conditions (suitable planet, etc.), the probabilities that life evolves are significant. But beware, this excludes animals, plants, and all eukaryotic organisms, that is, all those beings derived from a symbiotic association between different prokaryotic cells. This process, which occurred on Earth, will be the subject of the next section.

Before proceeding further, however, there is one last consideration. We have seen that, on a planet with suitable characteristics, according to our model, life establishes itself quickly and with a significant probability. All the descendants of the protocell that we have followed will have the same genetic code.

We remember that, on Earth, *All* living beings descend from a common ancestor, the famous LUCA (Fig. 11.1) we have already discussed. In fact, they all possess the same nucleotides and the same genetic code, the same essential amino acids, and all synthesize their proteins from the same type of structures, the ribosomes. The model based on a common ancestor is much more likely than schemes involving multiple origins.

But is this the general case, or is it possible that, on a particular star, life may appear independently more than once in different locations or from different situations? The model used does not absolutely oppose such a hypothesis. Nevertheless, it must be kept in mind that once a living being endowed with the appropriate characteristics emerges, it will begin to reproduce, evolving according to the rules of biological evolution and colonizing all available environments. In this way, it begins to consume all resources, particularly the biological molecules present, subtracting them from other potential processes of life origination that will not, therefore, be able to occur.

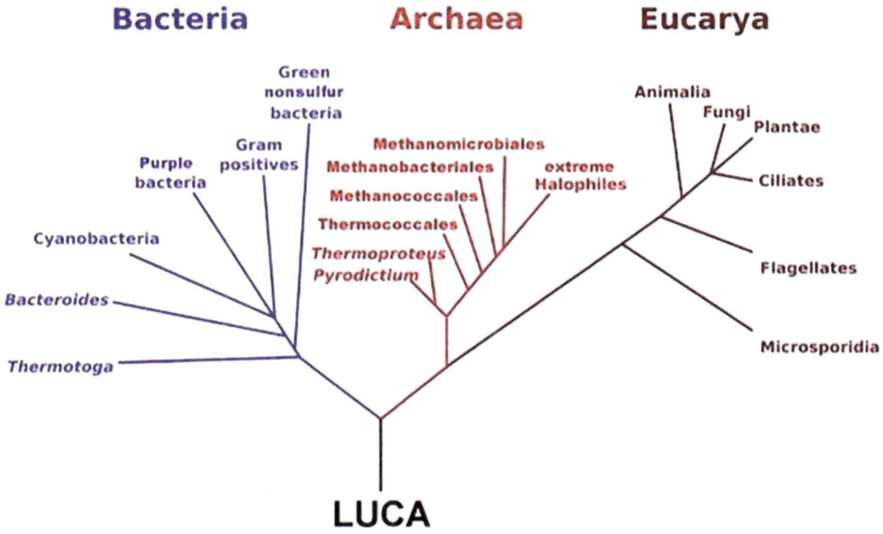

Fig. 11.1 Phylogenetic tree that links all the main groups, bacteria, archaea, and eukaryotes, to LUCA **(NASA)**

However, it is theoretically possible that life takes shape from two separate locations on the planet. In any case, the descendants of the first strain that appeared, or those of the fastest to spread and colonize the planet, will sooner or later enter into competition with those of the second. Then, they will tend to outcompete and make the second strain disappear. It is likely that in a few million years, the descendants of the first group will be the sole inhabitants of the planet, or they will have relegated the others to absolutely marginal roles.

We will see, in fact, the importance that for subsequent phases, the living beings that interact on the planet in question are *All* descendants of a single progenitor and therefore have all the same main characteristics, in particular, the same genetic code.

12

Fifth Drake: The Probability of Intelligent Life fi

In the fifth parameter, given the complexity, we immediately find ourselves having to divide the process into at least three large macrointervals:

Macrointerval A the appearance of the eukaryotic cell
Macrointerval B the appearance of animals (the metazoans)
Macrointerval C the appearance of intelligent civilization (ETC)

Each macrointerval will be divided into several phases, as was done for the fourth parameter. The substantial difference lies in the temporal scales of catastrophes: given that life, once formed, is decidedly more resistant compared to the biochemical environments that preceded it, the typical times of microcatastrophes $\Delta\mathbf{T}_j$ are now on the order of **100,000** years, while those of macrocatastrophes $\Delta\mathbf{T}_{0j}$ are on the order of **half a billion** years. The phases that divide the three macrointervals are as follows:

Macrointerval A

- the evolution of an aerobic bacterium
- the host-symbiont encounter
- the formation of pores and the extrusion of cytoplasmic extensions
- the "wrapping" of the symbionts and the disappearance of the host's cell wall
- the "penetration" of symbionts into the cytoplasm
- the migration of DNA from the symbiont's genome to that of the host
- the acquisition of the eukaryotic cytoplasmic membrane
- the incorporation into a single coating and phagocytosis

E. Mieli et al., *The Living Galaxy*, https://doi.org/10.1007/978-3-031-67324-5_12

Macrointerval B

- the acquisition of a complex life cycle
- the aggregation of zoospores and the formation of the synzoospore
- the sedentary colony composed of differentiated cells
- the production of collagen

Macrointerval C

- the increase in size of metazoans (with the acquisition of the nervous and vascular systems)
- the development of limbs
- the conquest of the mainland
- the differentiation of terrestrial animals
- the acquisition of sociality
- the acquisition of upright stance and manual dexterity
- the change in diet and the growth of the brain
- the organization of the brain for abstract thought
- the birth of articulated language and technique

12.1 The Starting Point; The Conditions of Stability of a Planet

For life as we know it to be sustainable, water must be present, especially in its liquid form. This substance can be broken down into its constituent elements, hydrogen and oxygen, by ionizing radiation from the sun. To block this radiation, a protective layer like ozone ($\mathbf{O_3}$) must form in the atmosphere. This molecular form is the product of common oxygen ($\mathbf{O_2}$) reacting to the main solar electromagnetic rays, which are thus shielded.

Paradoxically, the oxygen resulting from the decomposition of water does not contribute to the formation of the ozone layer, because the phenomenon of hydrolysis is very slow and the released gas is sequestered by the oxidation of minerals present in the environment. To form $\mathbf{O_3}$, oxygen molecules must be able to remain in the atmosphere for a sufficient time; for this, their production must be regular and abundant. The only process that allows an abundant and regular production of this gas is photosynthesis carried out by chlorophyll-containing plants and cyanobacteria [95]. But producing oxygen is not enough; the planet must have enough mass to retain the oxygen molecules, without them being dispersed into space. A sufficient planetary mass is precisely one of the initial conditions evoked in the previous section of this work related to the fourth parameter.

We will return to this issue later, but remember that if water were to disappear, no form of life would be possible and, for this reason, it is necessary to stop its hydrolysis before it is too late. However, if these conditions were sufficient for the onset of life, we will see that they are no longer enough for the conditions we want to address now. In fact, we must ensure that the very existence of the planet is long enough to ensure biological evolution reaches the level that interests us. Now, from what we have seen for the third parameter, such times are measured in several billion years; the age of the Earth is indeed **4.54 ± 0.05 billion** years. At the end of the section, we will summarize the information that comes to us from the third parameter onwards to establish the number of planets stable enough for the development of life at every stage, from the most primitive to galactic civilization.

13

Macrointerval A: The Crucial Transition; The Onset of the Eukaryotic Cell

Current prokaryotes include archaea and all bacteria. They are microscopic organisms, significantly smaller than eukaryotes: despite the existence of “giant” bacteria, organisms with eukaryotic cells have on average a volume **15,000** times greater than prokaryotes. These latter, however, despite being smaller, are numerous and widespread everywhere: they represent the most substantial part of terrestrial biodiversity [31, 83].

The diversity of prokaryotes is not based on morphology, but on their metabolism. The set of metabolic differences that plants, animals, fungi, and all other eukaryotic organisms display is nothing compared to the array of different processes presented by bacteria and archaea. These latter were once simply considered prokaryotes specialized in colonizing extreme environments, where living conditions are (according to our standards) if not impossible, at least very difficult (hot springs at high temperatures, hypersaline ecosystems, anoxic environments). In reality, archaea seem to be present in the majority of environments existing on our planet [6].

On the other hand, bacteria are the organisms that present the most significant metabolic diversity. Some are capable of coexisting with hyperthermophilic archaea, while others perform the same type of photosynthesis carried out by chlorophyll-containing plants, with the release of $\mathbf{O_2}$. There are bacteria that cannot tolerate this gas, while others grow very well in its presence. There are even some that are capable of producing the energy they need thanks to the reduction of uranium. Finally, to underline the surprising capabilities of some representatives of this group, let’s remember *Rubrobacter radiotolerans* [22], one of the organisms most resistant to gamma radiation: it can tolerate doses thousands of times higher than those needed to kill a human. It seems,

E. Mieli et al., *The Living Galaxy*, https://doi.org/10.1007/978-3-031-67324-5_13

moreover, that there are various microbial strains, belonging to both bacteria and archaea, that present high tolerances to radiation.

Eukaryotes cannot compete with prokaryotes on their own ground. However, these organisms, thanks to their abilities and properties, occupy ecological niches precluded to bacteria and archaea. The eukaryotic cell differs from the prokaryotic one (Fig. 13.1) due to the presence of peculiar characteristics:

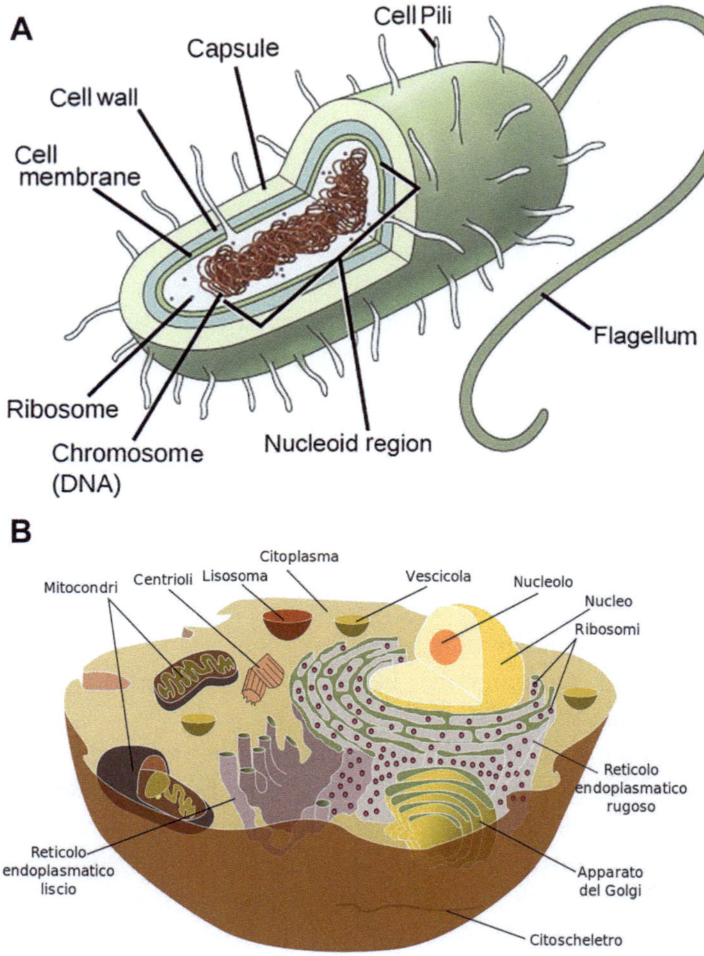

Fig. 13.1 5° Drake—macrointerval **A**: Comparison between a prokaryotic cell (**a**) and an animal eukaryotic cell (**b**), with the typical elements of the two cells indicated on the figure. The main differences between bacteria and archaea lie at the level of the cell membrane. The two cells are not to scale: the prokaryotic cell is about the size of the mitochondria indicated in the eukaryotic cell **((a) CNX OpenStax and (b) Giac83)**

- It has a **nucleus** with a double membrane, in which (almost all) the cellular DNA is contained, organized in the form of chromosomes;
- It contains **cellular organelles** with genetic material not present in the nucleus and internal membranes;
- It is equipped with a dynamic **cytoskeleton** (cellular skeleton) mainly made up of filaments constituted by the actin protein, which supports the membrane and allows its deformation (eukaryotes, originally, did not have a cell wall, although certain phylogenetic lines, such as plants, subsequently evolved one);
- Finally, they can develop a whole series of complex behaviors, including, among other things, **phagocytosis** (lost in eukaryotes with a cell wall), **sexuality** and aggregation into **multicellular organisms**.

A eukaryotic cell normally possesses a significantly larger amount of genes than a prokaryotic one: the largest bacterial genome contains **12** megabases [**Mbp**] (one base [**bp**] is a pair of paired nucleic acids; a megabase [**Mbp**] indicates a million of base pairs) of DNA, while the human one has **3000 Mbp**—and certain eukaryotes reach up to **100,000 Mbp** [60]. The ability to manage complex structures and processes, obtained thanks to a powerful collection of proteins that mediate their implementation, seems to be the great difference between eukaryotes and prokaryotes. In fact, theoretically, the various "eukaryotic" characteristics all seem to have their prokaryotic precursors, but the latter never take the decisive step towards complexity [27].

To better understand what this complexity consists of, let's see some typical examples. Let's start with sexuality; in the fossil record, the oldest known testimony is *Bangiomorpha pubescens* [11], a red algae, found in the sediments of the Canadian Arctic currently dated to just over **1 billion** years (**1 Gy**) [35, 54]. Although we do not know exactly whether the first eukaryotic organism to appear was capable of reproducing sexually, it seems clear that the oldest ancestor of all current eukaryotes was. No prokaryotic organism, on the other hand, has a sexual cycle, although some are capable of transmitting genetic material through horizontal gene transfer.

Regarding the constitution of multicellular organisms, although there are known bacteria capable of associating and forming filaments, no prokaryotic organism is capable of forming aggregates that can behave in such a coordinated way as to form a complete individual.

Finally, as already indicated at the beginning of the paragraph, unicellular eukaryotes are larger than prokaryotes, although there are exceptions. There are gigantic bacteria, but only a handful of forms are known in total. *Epulopiscium fishelsoni* and *Thiomargarita namibiensis*, for example, are true titans: their size makes them visible to the naked eye (in fact, they reach tenths of a millimeter), exceeding the size of many unicellular eukaryotic organisms! What allows these organisms to reach such a size? Their peculiarity is that they can have many copies (up to several thousand) of their genome relegated to the periphery, close to the cell wall, while most of the cytoplasm is metabolically inactive [76]. These adjustments would allow them to survive despite their excessive size. In any case, this strategy turns out to be an evolutionary *cul-de-sac*, because it does not translate into any complex behavior on the part of these prokaryotes.

Having said this, how to explain the differences between these two types of cells? How to justify the superior capabilities of eukaryotes compared to those of prokaryotes? Beware, these latter organisms are not at all inferior or less evolved than the former (just think about the fact that they present such a diversified range of metabolisms that they can be found absolutely everywhere). Simply, eukaryotes have evolved to occupy ecological niches precluded to bacteria and archaea. How did they do it?

13.1 The Eukaryotic Cell as a Symbiosis Between Prokaryotes

The answer would lie in the genesis of the eukaryotic cell, derived from a symbiotic association between prokaryotes and, more particularly, between an archaeon and a bacterium [4, 18, 71]. Some specialists even think that viruses were involved in the formation of the nucleus of the new type of cell. In any case, most authors now agree in considering eukaryotes as real "chimeras", obtained from more than one living being. But in order to obtain a symbiosis of this type—more precisely, an endosymbiosis, in which various symbionts live inside the host cell—it is necessary that the subjects involved in the process possess the same genetic code, in order to understand the same instructions contained in the genes. For this reason, in the previous paragraph, the importance of descent from the same ancestor in subsequent evolution was emphasized.

But let's get back to our main point. All current eukaryotes have organelles, limited by a double membrane, that derive from symbiotic prokaryotes: the **mitochondria** and the **plastids**. The latter are only present in photosynthetic organisms, such as green plants, as they are essential for such processes. The acquisition of these organelles was accompanied by a transfer of genes from the symbiont's genome to that of the host, a phenomenon that always occurs in ALL endosymbioses. The mitochondria, in particular, would have derived from a single form of bacterium belonging to the group of alpha-proteobacteria, which establishes the unique origin of this organelle. **It was precisely its acquisition that marked the birth of the eukaryotic cell** [72]. In fact, all current eukaryotes have mitochondria, except for the few forms that have transformed them or lost them, but still preserving some characteristic genes in their nucleus [118]. In any case, it is believed that the common ancestor of all eukaryotes had a mitochondrion.

But what advantage does this organelle confer? Is the fact of breathing $\mathbf{O_2}$ enough to justify the eukaryotic capabilities? Aerobic respiration (the oxidation—a real combustion—of nutrients by $\mathbf{O_2}$) is more advantageous than anaerobic, at least **6** times more. However, the presence of oxygen increases the costs of protein production (remember that living beings were formed in an anoxic environment, i.e., devoid of free oxygen). The presence of this gas increases **13** times the expenses of protein production, compared to its absence. Moreover, in various bacteria, aerobic metabolic processes develop much faster than in the mitochondrion. Therefore, the advantage does not seem to be linked only to the respiration of $\mathbf{O_2}$. In reality, the benefit conferred by this organelle lies in the ability to greatly increase the **energy available per gene**: with the term "energy per gene" we mean the cost necessary for *gene expression*: the cost to produce proteins and other cellular components. By increasing the energy per gene—and the presence of mitochondria allows the eukaryotic cell an increase between **4** and **6** orders of magnitude—the amount of energy that can be devoted to gene expression increases, thus also increasing the number of genes that a cell can manage. And the more genes there are, the more the cell becomes capable of "complex" processes and behaviors. Remember that the eukaryotic genome includes a much higher number of genes than the prokaryotic one.

Let's try to quantify better what has been said. Starting from experimental bases, it is possible to evaluate the average metabolic rate of a bacterium in normal growth: its value is about **0.19 W/g**. Since its mass is $\mathbf{2.6 \cdot 10^{-12}g}$, the total power of the cell will be given by the following formula:

$$0.19 \times 2.6 \cdot 10^{-12} = \mathbf{0.49 \cdot 10^{-12}\ W}$$

If we consider a protozoan (a unicellular eukaryotic being), we have an average metabolic rate of about **0.06 W/g**, but a much larger mass, of $\mathbf{4 \cdot 10^{-8}\ g}$. The total power of the cell will then be given by:

$$0.06 \times 4 \cdot 10^{-8} = \mathbf{2.400 \cdot 10^{-12}\,W}$$

Starting from these values, knowing the number of genes contained in the DNA of the bacterium (**5000**) and of the protozoan in question (**20,000**), we can calculate the energy available per gene for the two organisms:

$$\textbf{–bacterium}\; 0.49 \cdot 10^{-12} / 5000 \quad \cong \mathbf{10^{-16}\,W / gene}$$
$$\textbf{–protozoan}\; 2.400 \cdot 10^{-12} / 20{,}000 = \mathbf{1.200 \cdot 10^{-16}\,W / gene}$$

It is easy to notice that, in this particular case, the eukaryotic cell has an energy available per gene that is about **1200** times higher than that of the prokaryotic cell. Using the metabolic rates, masses and genome sizes of other organisms, even larger values can be found for eukaryotic organisms.

It is not size that explains the superior performance of eukaryotes, but rather their capabilities that allow them to reach certain dimensions. Let's give a practical example to better simplify the difference between prokaryotic and eukaryotic organisms. Let's imagine that, for a wedding lunch, you want to prepare tortellini with a particular meat filling and a special sauce for **30** guests. Who will be more efficient? A team of three specialized chefs (one for the tortellini dough, another for the meat filling, and the last for the sauce) or three chefs each preparing the same entire dish for 10 guests? It is easy to realize that the team of three people, in which each phase of the work is carried out separately (decentralization), will produce more quickly and effectively, that is, with less waste, a large number of portions compared to asking several chefs to do all the work. The statement remains valid even if we considered many more chefs (10, for example) each doing all the work. In fact, the division and specialization of labor are characteristics that allow optimizing the results.

Let's remember the giant bacteria, with various copies of their genome distributed near the cell membrane and the central part of the inert cytoplasm. To be able to produce more energy, a larger effective surface is needed (in the case of bacteria, it is the entire surface of the cell membrane). But a larger surface also requires more genes to manage and manufacture it (in the case of bacteria, the entire genetic heritage, consisting of a single chromosome, is duplicated). However, all these genes require energy to be managed. Moreover, as the surface increases, the cytoplasmic volume also increases, which also requires energy to be managed. Giant prokaryotes have found a solution to increase their size, but essentially, the energy available per gene does not change compared to that available for a standard-sized bacterium. Therefore, from an evolutionary point of view, it is not an improvement, quite the contrary...

Instead, the eukaryotic cell does increase its size, but it does not have n copies of the entire genome: it decentralizes energy production in the various mitochondria (each of which uses the effective surface of the membrane that delimits it), which, in turn, transfer their genome into that of the nucleus, except for the genes strictly necessary for the control of the functioning of the energy production redox chain (it is estimated that mitochondria retain only about **1%** of their original estimated genome) [40]. In this way, the various components of the eukaryotic cell divide the tasks: the nucleus preserves the genome and replicates it, the cytoplasm reserves the production of cellular material, and the mitochondria, which no longer need to deal with protein synthesis, dedicate themselves exclusively to energy production, increasing efficiency and benefiting the entire organism. In this case, the increase in energy due to the increase in the effective production surface (the set of all mitochondrial membranes) does not have to be subtracted a part equivalent due to the multiplication of the genetic code: one copy of this is enough, properly managed, inside the nucleus (plus some genes necessary for the management of the redox chain in each mitochondrion). The energy available per gene is effectively larger for this type of organism!

To recap: eukaryotes derive from a symbiotic association between an archaeon and a bacterium (the latter will become the mitochondrion). Currently, it is thought that the symbiotic bacterium was a facultative aerobe, capable of breathing $\mathbf{O_2}$ when present, but also equipped with anaerobic metabolism in the absence of this gas. The consumption of $\mathbf{O_2}$, therefore, is present from the first steps of the onset of eukaryotes, and this fact underlines the importance of this gas in the evolution of living beings. In this regard, it is enough to remember what we have already said about the ozone layer ($\mathbf{O_3}$), which protects us from ionizing radiation (enough to separate electrons from atoms) coming from space and allows us to live on land. Without free oxygen, not only would water be decomposed into its fundamental elements, but the evolution of living beings would have been deeply disrupted. In fact, the level of $\mathbf{O_2}$ influences the synthesis of **cholesterol** [107], an indispensable molecule present in the cell membrane of ALL animals. Not surprisingly, the onset of the eukaryotic cell is posterior to the *Great Oxidation Event* (GOE), which occurred around **2.4** and **2.1 billion** years ago [64]. This event signals the presence of free $\mathbf{O_2}$ in the environment, as evidenced by the geological formation of layers rich in iron oxides, the *Red Beds*. In short, the importance of this gas for the first eukaryotes and their descendants is no longer to be demonstrated.

Although there are some rare cases of bacteria present inside other prokaryotes, the symbiotic association that allows the emergence of eukaryotes is considered a **unique event** (although a recently discovered organism has raised the question) or **extremely unlikely**. Indeed, it is noted that the emergence of eukaryotes, placed between **2.1 billion** years ago and **1.6 billion** years ago [49], occurs after more than **1.5 billion** years from the appearance of prokaryotes, which occurred around **3.7 billion** years ago, or even earlier [24, 89]. During this long period of time, eukaryotes are absent, while bacteria and archaea differentiate abundantly [99].

Despite this, we want to question, in this context, the possibility that the onset time of eukaryotes could have been influenced by different parameters compared to the presumed improbability of the symbiotic phenomenon. For example, one of the key factors could simply be the time required to reach a certain level of $\mathbf{O_2}$ in the environment, without which the association would not have occurred.

But why, then, do all eukaryotes descend from a single ancestor and, subsequently, no other symbioses of this type have been recorded, despite there being a relative abundance of $\mathbf{O_2}$ in the environment? We do not have a definitive answer to this question. The only one that comes to mind is that, having worked perfectly the first time, eukaryotes gradually occupied all the available niches, limiting competition and preventing the phenomenon from repeating or strongly limiting it (remember what was said about the transition from non-living to living and the descent of all current living beings from a single ancestor).

13.2 The Inside-Out Model for the Emergence of the Eukaryotic Cell

Due to the presence of a rigid cell wall, prokaryotes (whether they are bacteria or archaea) are not very suitable for being "colonized" by other microorganisms. So how to solve the problem of the association of an archaeon with endosymbiotic bacteria?

A recent model, called *Inside-out*, could provide an interesting and original answer to the problem of a symbiotic association between an archaeon and a bacterium and would allow explaining in more detail the emergence of the eukaryotic cell.

The authors Baum and Baum [7, 8] propose that it is not the bacteria that penetrate inside the cell wall of the archaeon, but that the latter is able to produce, through specific pores homologous to those of the nuclear membrane of eukaryotes, cytoplasmic outgrowths, which, over time, can wrap around the associated microorganisms stationed on its wall (Fig. 13.2a).

According to this model, the outgrowths would cover the entire organism, wrapping around the archaeon cell (Fig. 13.2b), which would in fact become

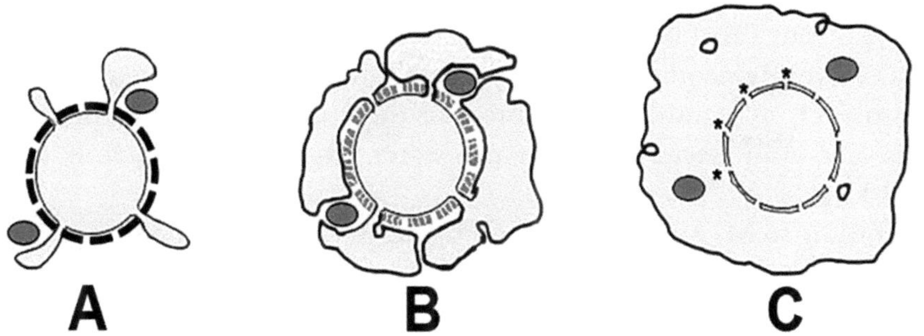

Fig. 13.2 Simplified diagram of the *Inside-out* model: (**a**) the protrusions emerge from pores in the wall (thick black line) of the archaeon; (**b**) the protrusions develop and begin to wrap around the associated bacteria (the two gray ovals), while the wall begins to fade; (**c**) the cytoplasm completely covers the archaeon and the bacteria have penetrated through the membrane, finding themselves inside: the original structure of the archaeon, covered by a double membrane with pores (four of these are marked with asterisks), will become the nucleus of the new cell **(IJA 14/06/2023 Mieli, Valli, Maccone)**

the "nucleus" of the new organism, covered by a double-layered membrane derived from the original one plus the one added by the lobes of the outgrowths. Gradually, the archaeon's cell wall would also disappear, having become superfluous. Finally, the associated bacteria would lose their cell wall to find themselves also in the cytoplasm protected by a double membrane (Fig. 13.2c). Essentially, the symbionts would not enter the host, but it would be the host's cytoplasm that has come out to cover them.

Certain archaea are perfectly capable of generating extracellular protrusions [70]. Moreover, as we will see later, the Inside-out model is perfectly compatible with the hypotheses made about the possible nature of the host. For this reason, we will adopt it to describe the steps leading to the formation of the eukaryotic cell.

13.3 The Onset of the Eukaryotic Cell, Phase by Phase. The Starting Point; The Release of Oxygen and Its Diffusion in the Environment

We have seen the importance of O_2 concentration for the onset of eukaryotic cells and the evolution towards intelligent life forms. The only process that ensures the production of significant quantities of this gas is photosynthesis, a purely biological phenomenon. But then, when do the first organisms capable of releasing O_2 thanks to the process of photosynthesis evolve? Remember that only some photosynthetic organisms are able to release O_2 as a result of this process: these are chlorophyll-containing plants and cyanobacteria, which use water, H_2O, as an electron donor (Fig. 13.3).

According to M. Ageno, the first living organisms were not only capable of photosynthesis, but they even released O_2. In fact, the Italian physicist argues that electrons logically had to be obtained from a very common substance in nature, water or H_2O, precisely. For this type of photosynthesis, special complex pigments, chlorophylls, are needed. M. Ageno reports that laboratory experiments like those carried out by Stanley Miller have shown that the abiotic synthesis of such molecules is perfectly possible.

Fig. 13.3 Release of oxygen from the first photosynthetic cyanobacteria in shallow depths during the GOE **(2.4–2.1 Gy)** Powered by OpenAI

According to other specialists, however, the first living beings were chemo-autotrophs. Although the GOE, occurred between **2.4** and **2.1 billion** years ago, experts believe that prokaryotes capable of producing $\mathbf{O_2}$ (cyanobacteria) were present on our planet at least from **3 billion** years ago, although certain clues, not always decisive, suggest that they could have existed much earlier.

However, there is no contradiction between these dates, as the process of accumulation of $\mathbf{O_2}$ in the environment is considered a slow process and opposed by other phenomena. It remains that the content of this gas in the environment became equal to $\mathbf{10^{-2}}$ times approximately the current value only after **2.5 billion** years, manifesting with the ability to produce deposits called Red Beds. Note that the appearance of eukaryotes **is subsequent** to this event.

13.4 The First Phase; The Evolution of an Aerobic Bacterium

The current theory of the onset of the eukaryotic cell involves a symbiotic association between an archaeon, the host, and various individuals of a bacterial strain, the symbionts. William Martin and Miklós Müller in 1998 [73] hypothesized that the host archaeon was an anaerobe, strictly dependent on $\mathbf{H_2}$, while the symbiotic bacteria were able to breathe $\mathbf{O_2}$ and produced $\mathbf{H_2}$ as a metabolic waste product. Such behaviors would have ensured complementarity between the organisms (Fig. 13.4).

Therefore, an aerobic bacterium was necessary, at least partially. But when did such organisms first appear? Research on Precambrian layers has made much progress in recent years and has documented an extraordinary variety of organisms during the first billions of years of our planet. Not only that, it has been documented, around **3.4 billion** years ago, a complex ecosystem, including microorganisms capable of producing hydrogen sulfide ($\mathbf{H_2S}$) plus other organisms, builders of stromatolites, dependent on this substance which they used as an electron donor to perform photosynthesis.

As you can see, even in ancient times, prokaryotes had differentiated and had formed complex communities where various complementary ecological niches were occupied. With such assumptions, given the metabolic versatility of bacteria, it is reasonable to think that, once $\mathbf{O_2}$ became available in the environment, some microbial strain became capable of exploiting this gas as a resource to produce energy in a few thousand years at most: the probability is estimated between 0.4 and 0.6 every 2000 years, with a micro-catastrophe time limit of 100,000 years.

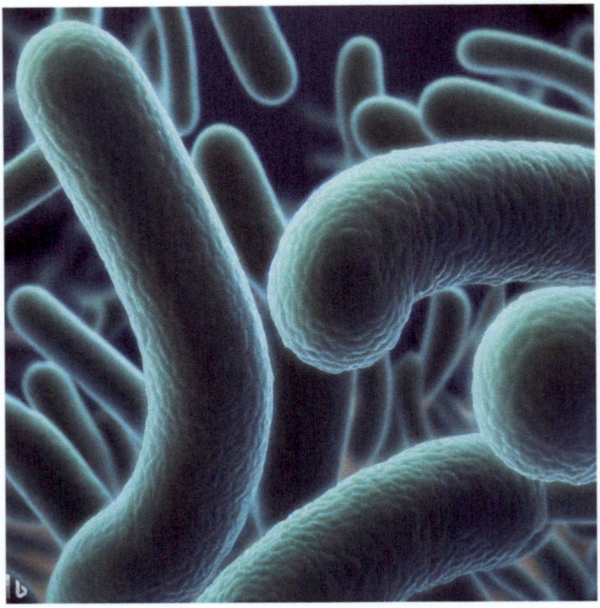

Fig. 13.4 Formations of the first aerobic prokaryotes during the GOE (2.4–2.1 Gy) Powered by OpenAI

PHASE 1

a_1	b_1	ΔT_1	ΔT_{01}
0.4	0.6	2000	100,000

13.5 The Second Phase: The Host-Symbiont Encounter

Let's now focus on the host organism. Currently, thanks to a series of protein homologies, it is believed that eukaryotes share a common ancestor with the group of archaea called "Asgard archaea" or even that such ancestor was directly within this group [25, 63, 104].

The Asgard archaea constitute a super-phylum known mainly thanks to genetic material found in the environment. Despite this, they turn out to be very widespread organisms, whose remains have been found in marine, lake and terrestrial sediments. They are mostly anaerobic and mainly spread in hydrothermal springs and/or methane-rich areas. One strain would even be anaerobic and dependent on $\mathbf{H_2}$, the requirements requested by Martin and Müller in 1998 for the host of the symbiotic association. They are therefore the ideal organisms to search for the archaeon protagonist of the onset of the eukaryotic cell.

Recently, moreover, a particular microorganism has been described, capable of surviving only thanks to a symbiosis with other microbes, which belongs to a phylum of the group of Asgard archaea, that of Lokiarchaeota (Fig. 13.5), considered close to eukaryotes. The described prokaryote—this time it is not just about simple analysis of gene sequences found in the environment, but the identification of a real existing organism—has the ability to generate cytoplasmic outgrowths [46]. This ability allows it to facilitate the exchange of material with external symbionts. Therefore, in light of all these facts, our approach consisting in the use of the Inside-out model appears legitimate.

Although the origin of the Asgard archaea is not known, from their supposed diversity it should be old enough. We consider, therefore, that the group already existed at the time of the appearance of aerobic bacteria, even if, being made up of mostly anaerobic prokaryotes, probably, the two types of organisms initially lived in different environments. However, currently, it is believed that the symbiont bacterium was a facultative aerobe, capable of populating the environments frequented by anaerobic archaea.

We therefore consider that an interval of 10,000 years is more than sufficient for the encounter to take place and for the association to form (associations between archaea and bacteria also occur currently, although the latter remain outside the former): the probability is estimated between 0.01 and 0.02 every 10,000 years with a micro-catastrophe time limit of 100,000 years.

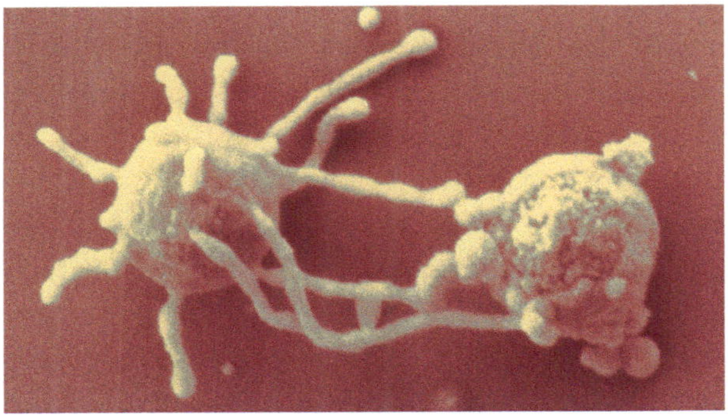

Fig. 13.5 Lokiarchaeota belonging to the set of archaea called "Asgard archaea", probable common ancestor with the eukaryotes (**Christa Schleper/Nature journal**)

PHASE 2

a_2	b_2	ΔT_2	ΔT_{02}
0.01	0.02	10,000	100,000

13.6 The Third Phase; The Formation of Pores on the Membrane and the Extrusion of Cytoplasmic Extensions

In the previous stage, we saw that the ability to produce cytoplasmic outgrowths is a characteristic of some groups of archaea. To allow the extrusion of pores, it is necessary that specific structural proteins are produced, such as the molecules that form the rings of the nuclear pore complex of eukaryotes (**COPII**). Although prokaryotic equivalents of such molecules have been found, it seems that these proteins are not homologous to those of eukaryotes. It is therefore preferable to imagine an evolution *ex novo* in the host lineage, rather than assuming an older prokaryotic inheritance. That is, these structures do not derive from analogous molecules present in prokaryotes, but are novelties evolved in the strain that led to the emergence of eukaryotes.

In any case, since pores are necessary for protrusions and since this ability is widespread in various groups of archaea, it is logical to assume that a certain variety of support proteins have been produced, including those homologous to those of eukaryotes, from which the latter would have derived.

Once the pore is produced, the cytoplasm can extrude forming an outgrowth supported by cytoskeletal elements (Fig. 13.6). For these latter, some proteins homologous to those of eukaryotes have been found in most of the phyla belonging to the Asgard archaea. The formation of pores and outgrowths is treated as a single phase, due to the close link between the two processes (the first would make no sense without the second, whose probability can be set equal to 1, once the previous one has occurred).

The time required for the production of molecules homologous to those of eukaryotes to stabilize the pores and for the formation of those of the cytoskeleton is estimated in a few thousand years: the probability is estimated between 0.04 and 0.06 every 5000 years with a micro-catastrophe time limit of 100,000 years.

PHASE 3

a_3	b_3	ΔT_3	ΔT_{03}
0.04	0.06	5000	100,000

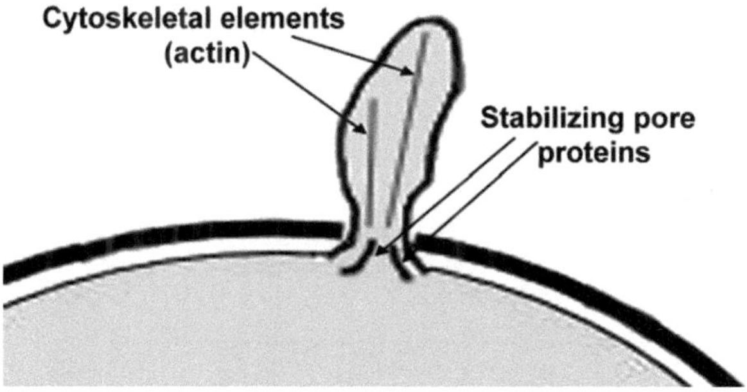

Fig. 13.6 Diagram of the host archaeo wall, with pore crossed by a cytoplasmic outgrowth, according to the *Inside-out* model **(IJA 14/06/2023 Mieli, Valli, Maccone)**

13.7 The Fourth Phase; The "Wrapping" of the Symbionts and the Disappearance of the Host's Cell Wall

The next step involves the development of cytoplasmic outgrowths that begin to approach and "wrap" the bacteria (Fig. 13.2b). These protrusions would have evolved to facilitate exchanges between the archaeon and the microorganisms in symbiosis with the former, which were on its external surface, being unable to penetrate the host's cell wall. The cytoplasmic expansions are supported by the cytoskeletal elements already evoked in the previous phase, which continue to develop. At the same time, the cell wall of the archaeon begins to regress, until it disappears completely, for two reasons:

- It becomes essentially useless, as it is covered by the cytoplasmic extensions, in contact with the internal environment of the archaeon;
- It is counterproductive because it hinders a greater extrusion of the cytoplasm.

At this point, due to the disappearance of the cell wall, the membrane of the extrusions, folded inward, overlaps the original membrane that covered the archaeon (Fig. 13.2c). Thus, the future nucleus of the new cell begins to take shape, consisting of the region formerly occupied by the archaeon and delimited by a double lipid membrane, equipped with pores.

The endoplasmic reticulum, a membrane with multiple folds located near the nucleus, where protein synthesis occurs, would derive from the folds of the membranes of the extrusions.

The duration of this phase is given by the wrapping of the host-symbiont complex by the outgrowths coming out of the pores and by the complete disappearance of the archaeon's cell wall; it can be estimated at a few thousand years: the probability is estimated between 0.01 and 0.02 every 2000 years with a micro-catastrophe time limit of 100,000 years.

PHASE 4

a_4	b_4	ΔT_4	ΔT_{04}
0.01	0.02	2000	100,000

13.8 The Fifth Phase; The "Penetration" of the Symbionts into the Cytoplasm

The following phase envisions that the bacteria are completely enclosed by the cytoplasmic expansions of the host cell, losing their cell wall to facilitate easier communication with the host's cytoplasm. Although it is difficult to establish the origins of mitochondria due to the minimal amount of genome they have retained, recent studies indicate that the ancestors of these organelles should be sought among the *alpha-proteobacteria*, and more specifically, among bacteria related to the *Rickettsia genus* [29] (Fig. 13.7).

These microbes are obligate endosymbionts that parasitize the cells of eukaryotes (mainly insects, but humans can also be infected), penetrating them thanks to their ability to lyse lipid membranes. However, the ancestors of mitochondria did not need to possess any penetration ability, as they were enveloped by the outward protruding lobes of the host cell. The disappearance of their cell wall is considered a step to enable greater exchange with the host's cytoplasm.

As Baum and Baum emphasized, this process absolutely does not involve phagocytosis (a phenomenon that, as we will see later, will only occur in the final stages). The bacteria were external symbionts, and they found themselves inside the host due to processes initiated by the latter. On their part, they merely lost their cell wall, remaining encased in a second membrane (relative to their own) derived from that of the host extrusion that enveloped them. This process is considered separate from the previous one, for the simple reason that we do not know if the disappearance of the two walls—that of the host and that of the symbionts—was simultaneous or not. In any case, **by the term "penetration of symbionts" into the cytoplasm, we only refer to the disappearance of the rigid cell wall of the symbionts.**

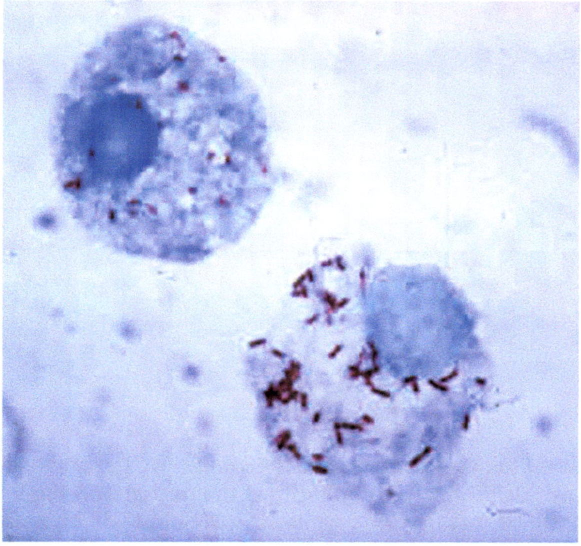

Fig. 13.7 *Rickettsia* is a small bacterium that grows inside the cells of its hosts

Considering the mechanisms of this process, the overall time expected for the completion of this phase is rapid, estimated at no more than a few decades. However, we attribute it a low probability of 0.01–0.02 every 50 years, with a microcatastrophe time limit of 100,000 years, to account for the risk of compromising the association following the change of conditions and the more direct relationship with the host's cytoplasm.

PHASE 5

a_5	b_5	ΔT_5	ΔT_{05}
0.01	0.02	50	100,000

13.9 The Sixth Phase; The Migration of DNA from the Genome of the Symbiont to That of the Host

Once endosymbiosis is established—with the symbionts now residing inside the host—a universal process is set in motion that occurs **every time these types of associations are produced**: the migration of genes from the genome of the symbionts to that of the host. When we talk about the "migration" of genes from the symbiont to the host, we mean their elimination from the

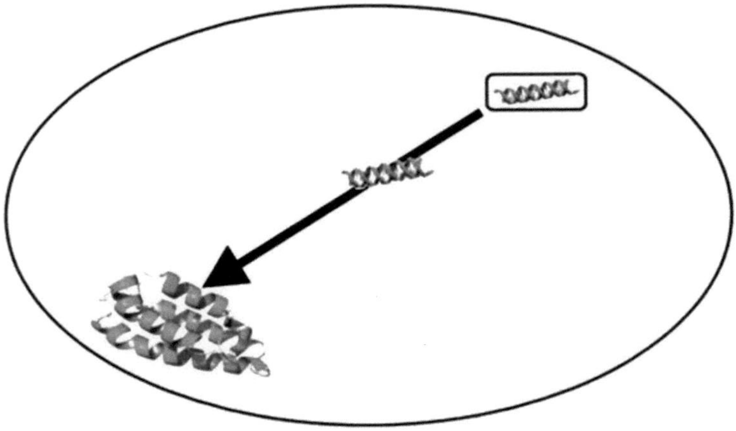

Fig. 13.8 Migration (**along the arrow**) of the symbiont's genome (**small rectangle at the top right**) towards the DNA of the host (**at the bottom, on the left**); the genome thus transferred will be lost from the symbiont's DNA but will be preserved in that of the host **(IJA 14/06/2023 Mieli, Valli, Maccone)**

genome of the former and their complete transfer to that of the latter (Fig. 13.8). This process allows the association to increase its efficiency, as the host takes care of protein synthesis, including that of various molecules of the symbionts, while the latter concentrate on their specific activities: energy production or the production of particular biological compounds.

The older the endosymbiosis, the more significant the amount of genetic material that has been transferred to the host's DNA. In this way, the two partners not only become complementary but also dependent on each other, especially the former symbiont. However, it is precisely by freeing the mitochondrion from tasks related to protein synthesis and allowing it to focus solely on energy production that enables the future eukaryotic cell to increase the energy available per gene, to levels unexpected for prokaryotes.

It is believed that currently, mitochondria retain only approximately **1%** of the genetic material possessed by their alpha-proteobacterial ancestor. According to current evidence, more than **1.5 billion** years (1,5 Gy) separates us from the symbiotic association that marked the onset of the eukaryotic cell.

How long must we wait for a sufficient migration of genetic material from the symbiont to the host? Certainly in the first organisms with sexual reproduction (the oldest known is *Bangiomorpha pubescens*, whose fossils have been found in sediments just over **1 Gy** old) the percentage of reduction must already have been comparable to the current one. Probably, however, a sufficient level had already been reached well before then. We can therefore estimate a timescale on the order of tens of thousands of years for this process:

with a probability between 0.1 and 0.2 every 10,000 years and a microcatastrophe time limit of 100,000 years.

PHASE 6

a_6	b_6	ΔT_6	ΔT_{06}
0.1	0.2	10,000	100,000

13.10 The Seventh Phase; The Acquisition of the Eukaryotic Cytoplasmic Membrane

The cellular membranes of archaea are constituted differently from those of bacteria and eukaryotes. These last two groups, unlike archaea, possess membranes made of the same types of lipids. This is a very important structural diversity between archaea on one side, and bacteria and eukaryotes on the other, because it involves building the membranes with slightly different materials and using different chemical bonds.

In fact, eukaryotic and bacterial cells possess membranes whose polar head groups are bound to lipid chains through ester-type bonds, while those of archaea use ether-type bonds. Moreover, the lipid chains of archaea are made up of branched, long isoprenoid alcohol molecules, while the chains of bacteria and eukaryotes are linear and unbranched (Figs. 13.9 and 13.10).

These differences are explained by the fact that the membranes of archaea are more resistant to the high temperatures where many of these organisms thrive (and where their ancestors likely evolved). But then, based on the inside-out model of eukaryogenesis, how can we explain the homologies between bacterial and eukaryotic membranes? What if the host is of archaeal nature and therefore has a different type of membrane structure?

Let's recall that in the previous phase, a large portion of the symbiotic bacterium's genome was transferred into the host's DNA.

This also applies to the genes responsible for assembling the bacterial membrane. Once they become part of the host's genomic heritage, biological evolution will push to achieve an "economical" situation: the simplest outcome is to produce only one type of membrane for all the constituents of the symbiotic association. But which one to choose—the original archaeal type or the imported bacterial type?

The answer can only be the second one. In fact, the membrane of the symbionts has specialized in efficient energy production through the oxidation of organic molecules by O_2, and this efficient energy production is precisely the key advantage conferred by these organisms to the symbiotic association.

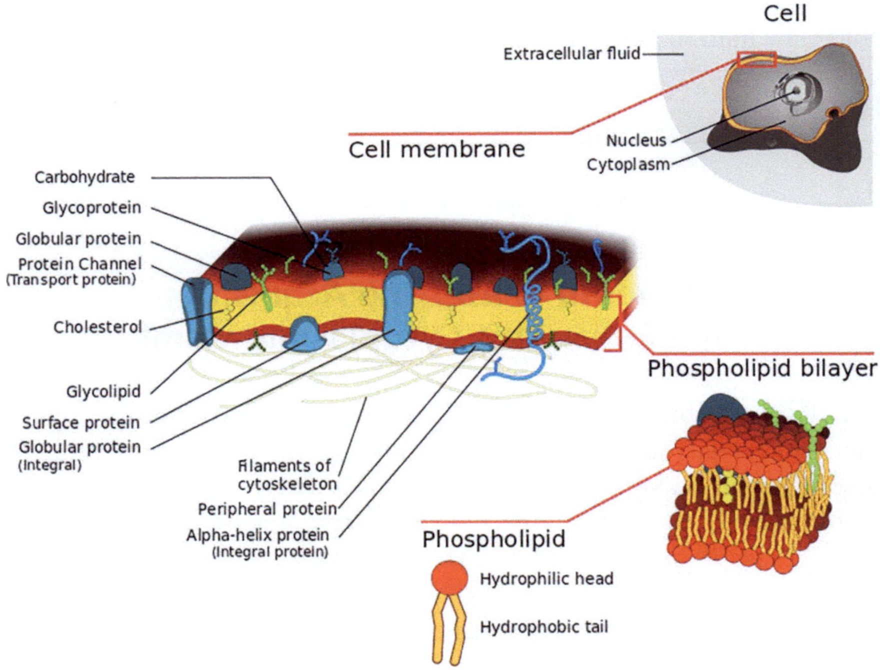

Fig. 13.9 Structure of the cell membrane (**Mariana Ruiz**)

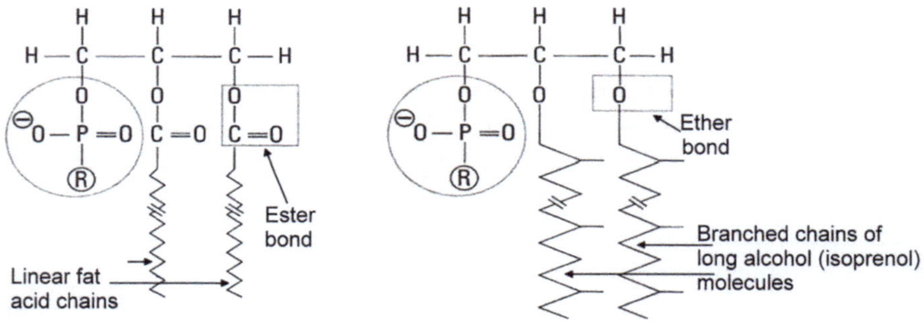

Fig. 13.10 Main differences between the constituents of the cell membranes of bacteria and eukaryotes (**on the left**) and those of the archaea (**on the right**): the polar head of bacteria and eukaryotes, binds to the hydrophobic chains through an ester bond, while archaea use ether bonds; the hydrophobic chains of bacteria and eukaryotes are made up of linear fatty acids, while those of archaea are formed by long branched alcoholic molecules (the various elements are not to scale with each other) **(IJA 14/06/2023 Mieli, Valli, Maccone)**

It follows that the only membrane whose nature can be modified is the original archaeal host membrane. In this way, the future eukaryotic cell will be covered by a membrane whose nature is typical of bacteria. The bacterial genes that determine the assembly of the cell membrane will begin to be selected as soon as they are fully integrated into the host's genome.

Thanks to the selective pressure exerted by evolution, this complete membrane replacement can take place in relatively short times—at most a couple of thousand years, with a probability estimated between 0.001 and 0.002 every 2000 years and a microcatastrophe time limit of 100,000 years.

PHASE 7

a_7	b_7	ΔT_7	ΔT_{07}
0.001	0.002	2000	100,000

13.11 The Eighth Phase; The Incorporation of the Host-Symbionts Ensemble into a Single Coating (Continuity of the Cytoplasm) and Phagocytosis

Finally, the lobes of the cytoplasmic expansions begin to come into contact and fuse with each other: the archaeal host's cytoplasm completely engulfs the endosymbiotic bacteria, exhibiting relative continuity among all its parts. Although the properties of the membrane components (mostly lipid molecules) favor fusion at the contact zones, somewhat as we have already seen in the third phase related to Drake's fourth parameter (transition from non-living to living), additional molecules are needed to complete this process. Proteins from the dynamin family, for example, are able to mediate the fission and fusion of biological membranes, allowing, among other things, the formation or fusion of vesicles. Many bacteria possess protein homologs of these eukaryotic molecules. It is therefore likely that eukaryotic dynamin-like proteins are derived from bacterial precursors, after the assimilation of the symbionts' genome by the host. The new contribution of these proteins, together with the cytoplasmic skeleton already discussed previously (third phase), also allows the onset of phagocytosis—the process that enables a deformable cell (therefore without a rigid wall) to ingest smaller solid objects (including cells smaller than itself). It is precisely because of the development of this complex protein system of bacterial origin that explains why the formation of the outer membrane and the appearance of phagocytosis only occurred in the terminal phases of eukaryogenesis.

Regarding other processes such as cell division or the acquisition of cellular cilia/flagella, these have evolved in parallel to those already described (or shortly after), integrating perfectly with the inside-out model. Further contributions to the nuclear genome could also have been obtained thanks to the action of viruses, acting in parallel to the phenomena thus far exposed. However, such a scenario does not alter the overall picture that has already been described.

The time predicted for the definitive transformation into a fully eukaryotic cell is estimated at a few thousand years: with a probability between 0.01 and 0.02 every 5000 years and a microcatastrophe time limit of 100,000 years.

PHASE 8

a_8	b_8	ΔT_8	ΔT_{08}
0.01	0.02	5000	100,000

At the end of the process we have just detailed, we can observe the emergence of a new type of cell, constituted by the symbiotic association between an archaeon that forms the nucleus (but let's not forget about the bacterial genetic contributions and, possibly, those transferred by viral means) also providing the cytoplasm of the organism, with a varying number of bacteria that have become mitochondria. Subsequently, the new eukaryotic organism begins to diversify, occupying ecological niches that were precluded to prokaryotes (which we have already discussed in the general section on the origins of eukaryotic cells) that imply an increase in size or the adoption of complex behaviors. Starting from **1 billion** years ago (**1 Gy**), but probably even earlier, we are certain that eukaryotes had acquired the ability to reproduce sexually. This latter property gives an acceleration to the biological evolution of the group and an increased ability in the production of biological innovations, as we will see in the following sections.

13.12 Evaluation of the Probabilities at the Passage of Each Stage

We have thus obtained the **32** input values to be inserted in step **1** of the calculation algorithm of the Maccone's lognormal statistical distribution (Table 13.1). We report in Fig. 13.11 the lognormal distribution of the macrointerval A process.

Table 13.1 Fifth Drake—macrointerval **A**: The **32** values of the frequencies $\mathbf{a_j}$ and $\mathbf{b_j}$ minimum and maximum, of the observation time $\mathbf{\Delta T_j}$ and of the microcatastrophe time $\mathbf{\Delta T_{0j}}$, for each phase described in the previous paragraph

Phase	Description	a_j	b_j	ΔT_j	ΔT_{0j}
1	The evolution of an aerobic bacterium	0.400	0.600	2000	100,000
2	The host-symbiont encounter	0.020	0.030	10,000	100,000
3	The formation of pores and the extrusion of cytoplasmic extensions	0.040	0.060	5000	100,000
4	The "wrapping" of the symbionts and the disappearance of the host's cell wall	0.010	0.020	2000	100,000
5	The penetration of the symbionts into the cytoplasm	0.100	0.200	5000	100,000
6	The migration of DNA from the symbiont's genome to that of the host	0.500	0.700	10,000	100,000
7	The acquisition of the eukaryotic cytoplasmic membrane	0.001	0.002	2000	100,000
8	The incorporation of the host symbiont set in a single coating (continuity of the cytoplasm) and phagocytosis	0.010	0.020	5000	100,000

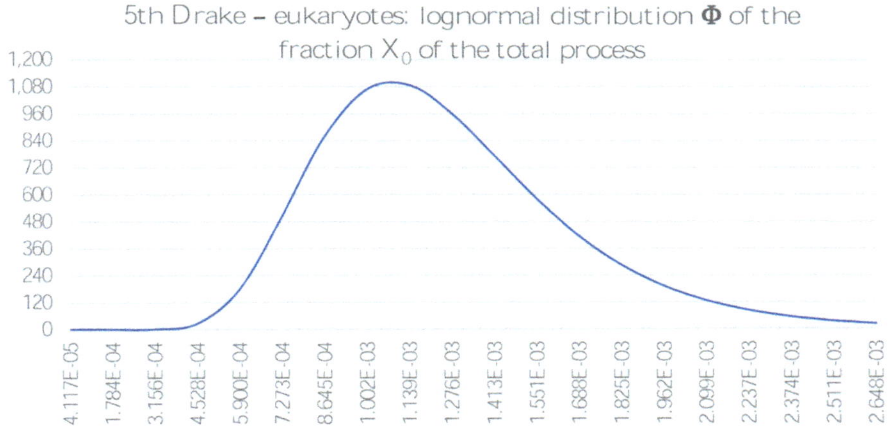

Fig. 13.11 Fifth Drake—macrointerval **A**: The lognormal distribution **Φ** of the process in the average period with $\mathbf{\langle X_0 \rangle = 1.26 \cdot 10^{-3}}$ **(IJA 14/06/2023 Mieli, Valli, Maccone)**

$\mathbf{\Delta T_0}$, as already written, is the sum of the $\mathbf{\Delta T_{0j}}$ and represents the average period equal to **800,000** years; while the long limit period $\mathbf{\Delta T}$ is set by us at about **500,000,000** years.

To conclude, at the end of our journey, we found, through the lognormal **Φ**, a probability of realizing a full cycle of the transition from non-living to living, in the average period $\mathbf{\Delta T_0 = 800,000}$ years, equal to about **0.126%,** (Fig. 13.11).

This value, applying the transformation rule of the probability described for the *phases* (*covariant with time*), that is:

$$\mathbf{p}_A = \mathbf{1} - (\mathbf{1} - \mathbf{p}_{A0})^n$$

translates, in the long term **ΔT = 500,000,000** years, into an average probability of the emergence of eukaryotes equal to:

$$\mathbf{f}_e = \mathbf{54\%}$$

between the two minimum and maximum values

$$\mathbf{f}_{e\,min} = \mathbf{29\%} \text{ and } \mathbf{f}_{e\,max} = \mathbf{71\%}.$$

The average gives us a value of about one case out of two, in a interval of **500 My**, comparable to that of the transition from non living to living. The phenomenon, therefore, does not seem to be so unlikely!

14

Macrointerval B: The Second Step; The Birth of Animals (the Metazoans)

In the previous section, we described a scenario that allows us to explain the emergence of the eukaryotic cell as a symbiotic association between prokaryotes. The new organism is not better or more evolved than the previous ones, but represents a higher level of complexity achieved by the different distribution of activities among distinct centers within the cell. For example, the mitochondrion is entrusted with energy production, being the organelle exempt from other major tasks. So, what will be the next step? It is possible to recognize, in the kingdom of life, an increase in complexity by considering step-by-step the biological entities obtained from the combination of those from the previous level. Let us explain this concept better: we have moved from **level I**, that of prokaryotes, to **level II** (the eukaryotic cell), by combining elements of the previous level (prokaryotic cells united in a symbiotic association). In the same way, it is possible to move to a further level, **level III**, by combining various elements of the previous level II, and so on [76]. This trend has already been rigorously suggested by Daniel W. McShea and Jean-Pierre Rospars [93]. For us, therefore, the next step will be sought in multicellular eukaryotes and, in particular, in the animal kingdom. In fact, in no other eukaryotic group or kingdom is it possible to encounter "intelligent" activities as in that of animals. Plants, evolutionarily speaking, while not at all inferior to their animal cousins, have chosen different solutions more suitable to their existence as immobile beings "rooted" in the soil. The same goes for fungi and other groups of eukaryotes.

But how are animals characterized? When do their traces appear among the fossils? Animals, better defined as *metazoans*, as we will call them later, have the following characteristics:

E. Mieli et al., *The Living Galaxy*, https://doi.org/10.1007/978-3-031-67324-5_14

1. They are **multicellular** eukaryotes, made up of differentiated cells;
2. They are **heterotrophs** (incapable of making their own food, they must find it in the environment in which they live);
3. They have a development that goes through very precise stages, including that of **embryo**;
4. They are capable of moving, at least in one of their different **life stages**;
5. Finally, as we will see better, ALL current animals, even the simplest ones, possess **collagen**, a structural element that intervenes in numerous processes (Fig. 14.1).

The oldest fossils of known metazoans date back between **630** and **550 million** years ago [43]. Meanwhile, between these dates and those of their first appearance (**2.1 billion** years ago), eukaryotes have differentiated and have already made most of their evolutionary conquests: just over **1 billion** years ago there are multicellular organisms equipped with sexuality and, shortly after, photosynthetic organisms appear that are the result of symbiotic associations between different eukaryotes. The same process is repeated, that is, the same process that led to the onset of the eukaryotic cell, but this time with eukaryotes as protagonists. In more or less coeval layers (around **1 billion** years ago), fossils of multicellular organisms having cells of at least two different types have been found. These eukaryotes, called *Bicellum brasieri*, are considered close to the group in which the ancestors of the metazoans are to be sought, based on their morphological characteristics. The curious thing is that they would not be marine organisms, but terrestrial ones.

But if multicellularity had already appeared before **1 billion** years ago, why do metazoans only manifest much later? First of all, let us point out that

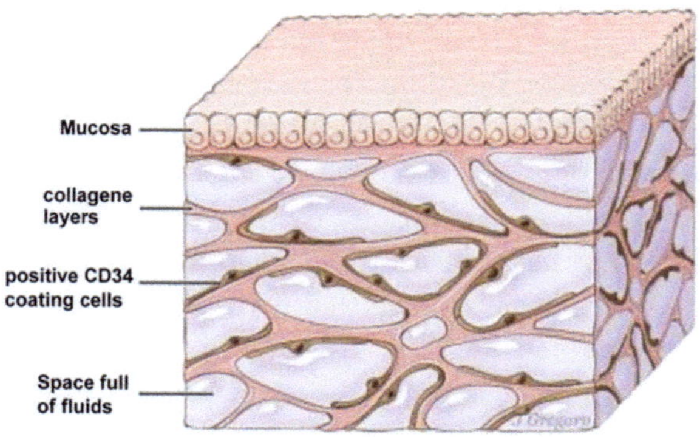

Fig. 14.1 Structure of cellular tissue held together by collagen or connective tissue **(Jill Gregory/MOUNT SINAI HEALTH SYSTEM)**

multicellularity has been achieved independently, and at different times, by at least **13** eukaryotic lineages, if not more [99]. This means that it is a property inherent to the condition of the eukaryotic cell.

However, it is not enough to be multicellular to be considered an animal. We remember that among the characteristics of modern metazoans is the ability to produce collagen. Now, the formation of this molecule requires an appropriate level of $\mathbf{O_2}$ in the environment. If we study the evolution of $\mathbf{O_2}$ levels, we discover new peaks and plateaus for values of $\mathbf{O_2}$ greater than those related to the Great Oxygenation Event (GOE) that we have already encountered. While during the Paleozoic there seem to have been values higher than the current one, in this context we are interested in the interval between approximately **0.8** and **0.5 billion** years ago. In this indicated period, there was a further increase in $\mathbf{O_2}$ compared to the values of the GOE. This event is referred to as the *Neoproterozoic Oxidation Event* (NOE): corresponding to it, $\mathbf{O_2}$ values comparable to the current ones are recorded [82].

The eukaryotic cell emerged among the periods in which these two different levels of $\mathbf{O_2}$ manifested. Soon after, eukaryotes became capable of performing photosynthesis thanks to the integration of a new endosymbiont, a cyanobacterium, which transformed into the **chloroplast** [13]. This new evolutionary step would have taken place between **1.5** and **1.2 billion** years ago. Subsequently, the new photosynthetic organisms and those that quickly evolved from them (among these eukaryotes are the distant precursors of terrestrial plants and their ancestors) will contribute, along with the already present cyanobacteria, to the oxygenation of the environment and will be fundamental for the achievement of the new peak of $\mathbf{O_2}$.

It is important to note the **correspondence** between the **NOE** (**0.8–0.5 billion** years ago) and the **onset of metazoans** (**0.63–0.55 billion** years ago) which leads us to suppose that, in order to evolve, animals required high levels of $\mathbf{O_2}$ in the environment. Why? Probably because otherwise they would not have been able to synthesize collagen, necessary to give the extracellular matrix and future tissues the required mechanical resistance. As we have seen in the case of cholesterol, a fundamental molecule for the eukaryotic cell membrane whose synthesis was favored by the GOE phenomenon, now the suspicion is again creeping in that the great conductor of the fifth parameter of Drake is once again oxygen whose net increase, first around **2.2 billion** years ago (GOE) and then at **0.6 billion** years ago (NOE), has allowed further levels of complexity in living matter. It would not be, therefore, about the improbability of the occurrence of other processes (or at least not all of them: let's remember the onset of the eukaryotic cell, seen at the end of Macrointerval A), but simply the time necessary to sufficiently oxygenate the planet!

Let's now move on to the search for the ancestors of metazoans and try to establish a model for their origins. The ***Holozoa*** constitute a group of eukaryotes that includes metazoans (but excludes fungi), in which various groups consist of unicellular organisms.

It is interesting to note that many animal proteins have homologs in eukaryotes of this group [78]. However, the organisms closest to metazoans are those that make up the choanoflagellates, flagellated unicellular organisms whose flagellum is surrounded by a distinct collar (Fig. 14.2).

These cells closely resemble choanocytes, the flagellated cells of sponges (simple sessile and filter-feeding animals, at the base of metazoans), which allow these organisms to convey nutrient particles towards their oral cavities. No fossils of choanoflagellates are known, but experts, applying the molecular clock, think that the group may have appeared between **1.05** and **0.80 billion** years ago, well before the onset of metazoans **around 0.63–0.55 billion** years ago [85].

Various models have been proposed to illustrate the origins of metazoans, but we will follow the *Synzoospore theory*, initially proposed by Alexey Zakhvatkin in 1949. Currently, the model is presented in the following form by Mikhailov [78]: a eukaryotic cell with a complex life cycle (including different morphological phases), produces spores that, instead of dispersing, join together to form a particular aggregate. This, in turn, transforms into a colony

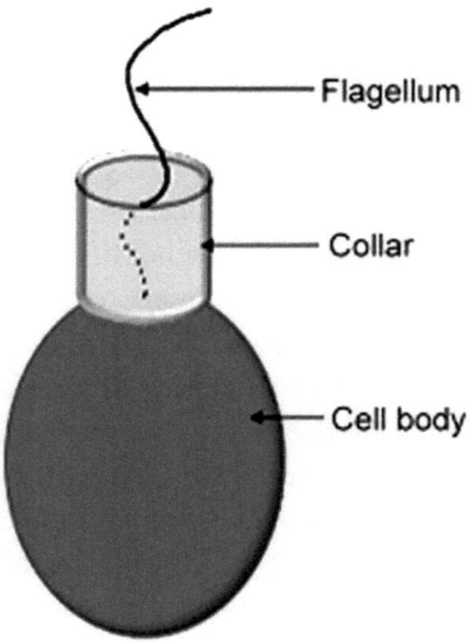

Fig. 14.2 Choanoflagellate—the flagellum (**dashed inside the collar**), the collar and the cellular body of the organism are indicated **(IJA 14/06/2023 Mieli, Valli, Maccone)**

with differentiated cells, which only reflect the different morphological phases of the initial eukaryotic cell (Fig. 14.3).

The basic idea is to start from a cell that presents different morphologies according to its life cycle (a phenomenon that occurs normally in various unicellular eukaryotes, including the choanoflagellates), whose zoospores, the cells derived from the **zygote** (the result of the fertilization of two sexual cells; Fig. 14.3a), join to form the **synzoospore** (Fig. 14.3b). Subsequently, each cell develops following a growth "desynchronized" with the others, so as to have a sedentary **colony** ("proto-larva") made up of different cells (even if all having the same genetic code, as derived from the same zygote), which settles on the bottom for a trophic **sedentary** phase (Fig. 14.3c).

Starting from this stage (and once the collagen that fills the extracellular matrix and provides mechanical support to the whole has been produced), evolution operates allowing the differentiation of the embryonic states that follow the *synzoospore*, in order to produce the first divisions and taxonomic differentiations within the group. Naturally, the final stage, to reproduce, generates a new zygote (Fig. 14.3: arrow between "**C** Sedentary colony" and "**A** zygote") and the cycle can start again.

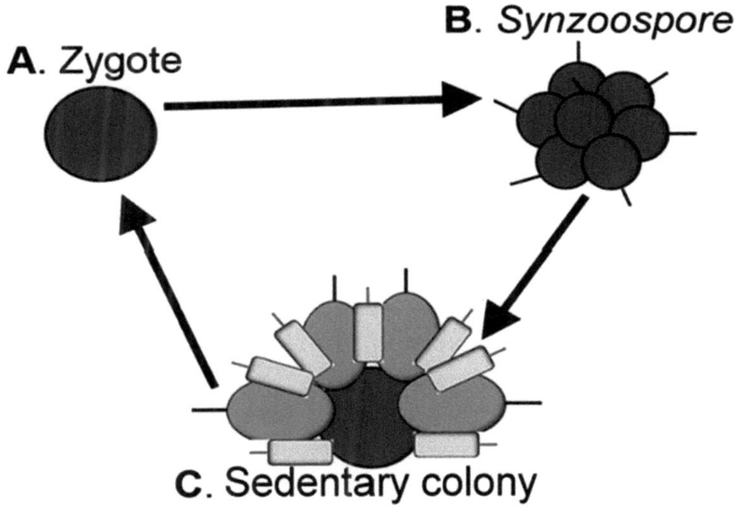

Fig. 14.3 The model of the *Synzoospore theory*: the zygote (**a**), dividing, produces the zoospores that join to form the *Synzoospore* (**b**); the asynchronous development of the cells that make it up produces a colony composed of different cells (even if all with the same basic genetic code) that acquires sedentary habits (**c**). To reproduce, the colony can generate a new zygote and the cycle begins again **(IJA 14/06/2023 Mieli, Valli, Maccone)**

14.1 The Onset of Metazoans, Phase by Phase. The Starting Point; The Choanoflagellates

These choanoflagellate organisms constitute the group of eukaryotes closest to that of metazoans, and it is therefore logical to search for the progenitors of animals among them. Choanoflagellates are all unicellular marine heterotrophic organisms that feed on bacteria. They predominantly reproduce asexually, but in at least one taxon, several genes related to the process of meiosis have been discovered, in turn linked to sexual reproduction. We also remember that the common ancestor to all current eukaryotes (LUCEA: *Last Universal Common Eukaryotic Ancestor*) is considered to have reproduced sexually.

Finally, we know that choanoflagellates share many genes with metazoans, and this allows us to consider them as the best candidates to find the ancestors of the metazoan group. Their genomic relatedness, similar collar cells, and inferred presence of sexual cycles make choanoflagellates prime candidates for the unicellular progenitors that multicellularized into the first animal forms.

14.2 The First Phase; The Acquisition of a Complex Life Cycle

The basis of the *Synzoospore theory* is that morphological diversity was acquired before multicellularity, in the sense that the organism that would eventually aggregate to form the first animal already had a complex life cycle comprising different morphological phases before associating and forming the multicellular entity.

We know that such a cycle exists in current choanoflagellates. But when did it evolve? Bearing in mind that it is assumed the group has existed for at least **800 million** years and that, probably, the first organisms could reproduce sexually (in any case, their ancestors could). Therefore, such a characteristic must have appeared quite early on. Among other advantages, being able to have different morphological types allows organisms to better cope with varying environmental conditions.

Given these considerations, the complex life cycle in the group could have evolved within a few tens of thousands of years, with a low frequency in the short term: the probability is estimated between 0.01 and 0.02 every 20,000 years with a mass extinction limit of 1,000,000 years.

PHASE 1

a_1	b_1	ΔT_1	ΔT_{01}
0.01	0.02	20,000	1,000,000

14.3 The Second Phase; The Aggregation of Zoospores and the Formation of the Synzoospore

Why should zoospores aggregate rather than go off on their own? At first glance, it would seem that the efficiency of dispersion could be reduced, but in reality this is not the case. Moreover, an increase in size can serve to deter potential predators from attacking the colony. Not only that, but this will make the colony more competitive when it comes to settling down to start a sedentary phase. This is, for example, the strategy used by many sponges. Let's not forget, then, that the colony (Fig. 14.3b) is formed by cells that possess the same genetic heritage (as they derive from the same zygote) and this favors intercellular communication and therefore the coordination of the whole, and that genes capable of producing cell adhesion substances are present within the group of Holozoa. However, the aggregation of the zoospores produces the *synzoospore*, the mobile phase of the organism's life cycle. It may also represent (it or its evolutionary development) the embryo of the future animal.

Knowing that there are choanoflagellates that can form colonies after cell division, this phase should also be able to be completed quickly (over the course of several thousand years): the probability is estimated between 0.02 and 0.03 every 15,000 years with a mass extinction limit of 1,000,000 years.

PHASE 2

a_2	b_2	ΔT_2	ΔT_{02}
0.02	0.03	15,000	1,000,000

14.4 The Third Phase; The Sedentary Colony Composed of Differentiated Cells

The asynchronous development of the different cells that make up the colony constitutes the third stage of the model. The morphological diversification of cells within the colony can be favored by the division of tasks, which allows increasing the efficiency of the organism (let's remember what we have already seen regarding the emergence of eukaryotes).

Having started from a unicellular organism characterized by a complex life cycle with different morphological stages, it should not be impossible to evolve a pool of genes that allows the asynchronous development of cells, so that some present a different morphology from the neighboring ones (morphology, however, provided by the general developmental plan of the organism; Fig. 14.3c). It is not about "inventing" anything new, except perhaps some regulatory genes. Let's remember, however, that metazoans share with their closest taxonomic groups not only structural genes but also others that regulate development, which can, through biological evolution, have given rise to those evoked above.

In this phase, the evolution of particular regulatory genes is necessary, starting from the pool present in the ancestors of the metazoans: the probability is estimated between 0.02 and 0.04 every 200,000 years with a mass extinction limit of 1,000,000 years.

PHASE 3

a_3	b_3	ΔT_3	ΔT_{03}
0.02	0.04	200,000	1,000,000

14.5 The Fourth Phase; The Production of Collagen

Starting from the previous phase, we already have a living being that we could almost define as a "metazoan". It is a heterotrophic organism (the zygote feeds on bacteria and, possibly, other food particles, before dividing) whose zoospores unite to form the *synzoospore*, the mobile phase of the life cycle. We also have a sedentary phase with cellular differentiation (but with cells that all possess the same genetic heritage). To reach the complete transformation into a modern animal, according to our definition, all that remains is the production of collagen (Fig. 14.4).

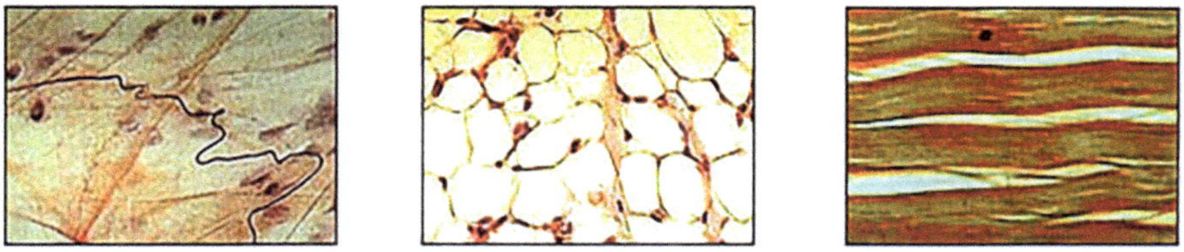

Fig. 14.4 Various types of connective tissue; from left to right: loose connective tissue, adipose tissue and compact connective tissue

We note that, regarding the onset of the embryo, the exact identification of this stage often depends on the animal group in question; we believe that the *synzoospore* or the early stages of existence of the sessile colony are a good representation. We underscore, finally, that a phase related to the production of collagen is not foreseen in the model discussed by Mikhailov and his colleagues nor in that of Sebé-Pedrós [99]. However, such synthesis becomes necessary to fill the spaces between one cell and another and to ensure mechanical resistance to the whole.

Apparently, collagen is absent in unicellular eukaryotes, but this situation should not surprise us, because in an organism made up of a single cell, the cytoskeleton is sufficient to give rigidity to the whole. Alternatively, the unicellular can choose other solutions, such as equipping itself with an exoskeleton (which will appear, however, only in more recent times). Collagen is therefore a substance that is produced *ex novo* by metazoans: there are no certain precursors among their unicellular ancestors.

However, for eukaryotic organisms that have sexual reproduction and differentiated cells, it should not be a problem to devise in a relatively short time a substance that can fill the interstices between one cell and another and provide the desired mechanical resistance: the probability is estimated between 0.01 and 0.02 every 50,000 years with a mass extinction limit of 1,000,000 years.

PHASE 4

a_4	b_4	ΔT_4	ΔT_{04}
0.01	0.02	50,000	1,000,000

Once all the basic characteristics that contribute to the definition of metazoans are gathered, biological evolution can favor the transformation of the cycle described above, developing some particular traits, which will lead to the differentiation of the main subgroups of animals. Let's remember, finally, the importance of the oxygen content in the environment: this may have been the determining factor that established the timing of the appearance of metazoans on our planet.

14.6 Evaluation of Probabilities at Each Stage

We have thus obtained the 16 input values to be inserted in step 1 of the calculation algorithm of the lognormal statistical distribution of Maccone (Table 14.1). The Fig. 14.5 shows the lognormal distribution related to the process of the macrointerval B in the medium term.

Reporting, with the method of the two previous cases, the probability of the metazoans $\mathbf{1.50 \cdot 10^{-2}}$ of the medium term $\mathbf{\Delta T_0}$ equal to **4 My** over the long term $\mathbf{\Delta T}$ equal to **500 My**, we obtain the probability of the macrointerval **B:**

$$\mathbf{f_m = 85\%}$$

Table 14.1 Fifth Drake—macrointerval **B**: The **16** values of the frequencies aj and b_j minimum and maximum, of the observation time $\mathbf{\Delta T_j}$ and of the microcatastrophe time $\mathbf{\Delta T_{0j}}$, for each phase described in the previous paragraph

		a_j	b_j	ΔT_j	ΔT_{0j}
1	The acquisition of a complex life cycle	0.01	0.02	20,000	1,000,000.00
2	The aggregation of the zoospores and the formation of the synzoospore	0.02	0.03	15,000	1,000,000.00
3	The sedentary colony composed of differentiated cells	0.02	0.04	200,000	1,000,000.00
4	The production of collagen	0.01	0.02	50,000	1,000,000.00

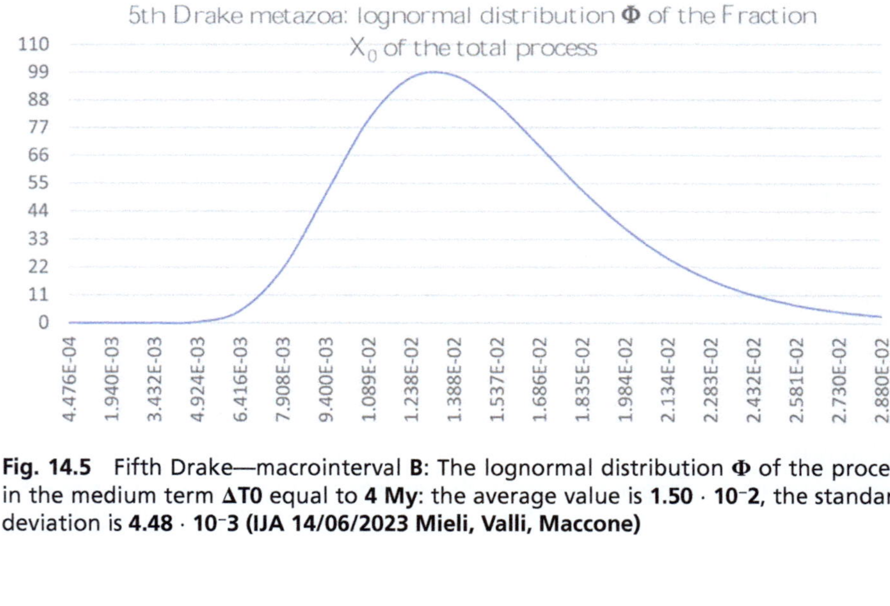

Fig. 14.5 Fifth Drake—macrointerval **B**: The lognormal distribution **Φ** of the process in the medium term **ΔT0** equal to **4 My**: the average value is $\mathbf{1.50 \cdot 10^{-2}}$, the standard deviation is $\mathbf{4.48 \cdot 10^{-3}}$ **(IJA 14/06/2023 Mieli, Valli, Maccone)**

between the two minimum and maximum values:

$$\mathbf{f}_{\mathrm{m\,min}} = \mathbf{60\%} \text{ and } \mathbf{f}_{\mathrm{m\,max}} = \mathbf{94\%}$$

In this case, the probability of carrying out Macrointerval B is decidedly high over half a billion years. The onset of animals, therefore, is a relatively easy process once the eukaryotic cell has evolved.

15

Macrointerval C: The "Solution" of Intelligence Deduced from the Definition of Kardashev, Focused on Energy per Individual, and Its Birth Within Metazoans (the *Homo* Case)

Biologists use various definitions of intelligence. For example, one of the most recent ones considers this faculty as an adaptive function that allows an individual to improve their behavior based on the context: the ability to modify behavior in the face of new or complex situations. It is clear that while this definition is useful for describing the "intelligent" behaviors of vertebrates and many other animals, for our problem, related to the fifth paragraph of Drake, much higher capabilities are required. And now we will explain why.

From a biological and social point of view, the definition of intelligence converges in the end on a single key concept: the onset of **abstract thought**, or the ability to combine different brain skills to construct new interpretive models of the environment and actions to be performed. This definition is absolutely appropriate for any aspect that is considered outside of one: the **amount of energy** that a species can manage through abstract thought itself. This is what we commonly call **technique** and that we can measure, primarily in terms of energy, as power in Watts (**W**) expressed per kilograms (**kg**) of body mass.

The new question we must ask ourselves is then: what power, per kg of body mass, does a certain species express? From simple empirical calculations we can deduce that living beings, particularly animals, develop an average basal chemical-metabolic power of about **1 W/kg** (we have seen previously, calculating the energy per gene, that this order of magnitude is true even for single prokaryotic and eukaryotic cells); this power is sufficient to support the animal in its usual biological activities. However, the current human civilization, thanks to the technique of fossil fuels and not only, for about a century, has a power estimated at **10 W/kg** which is a quantity ten times greater. This

E. Mieli et al., *The Living Galaxy*, https://doi.org/10.1007/978-3-031-67324-5_15

energy surplus allows us activities previously precluded such as, for example, the construction of large structures like missiles or telescopes; in other words, it makes us one of the potential civilizations of the galaxy. If we did not have this technique, as until two centuries ago, we could never hope to identify or be identified by other potential galactic civilizations, if not by pure chance.

According to the Russian physicist Nikolaj Semënovič Kardašëv, (from now on we will use the classic Anglo-Saxon transliteration of his name *Kardashev*) this parameter is fundamental for cataloging potentially communicative extraterrestrial civilizations (ETCs). For this purpose, in 1964, he devised a power scale on four main levels [51], later reviewed by Carl Sagan [94], according to the following recurring criterion:

$\mathbf{W_1 = 10^{16}}$ Watt	is the total solar power received by a rocky planet orbiting in its habitability zone
$\mathbf{W_2 = 10^{11} \cdot W_1}$	is the total power radiated by the star
$\mathbf{W_3 = 10^{11} \cdot W_2}$	is the total power radiated by the galaxy
$\mathbf{W_4 = 10^{11} \cdot W_3}$	is the total power radiated by the observable universe

As you can see, a recurring growth factor of $\mathbf{10^{11}}$ is respected between each level. This characteristic allows us to categorize a civilization with the simple formula:

$$\boldsymbol{K} = \frac{log_{10}\left(W_{ETC}\right) - 5}{11}$$

Where $\boldsymbol{K}$ is nothing more than the index at the base of the four levels $\mathbf{W_1}$, $\mathbf{W_2}$, $\mathbf{W_3}$ and $\mathbf{W_4}$, while $\mathbf{W_{CET}}$ is the power, expressed in Watts, harnessed by the civilization in its entirety.

On this scale, a civilization with $\mathbf{K = 1}$, or briefly K1, is able to manage an amount of energy equal to that provided to the planet by its own star. A K2 civilization manages all the energy of the star, a K3 civilization all the energy of the galaxy, and a K4 civilization all the energy of the observable universe. The current human civilization of about **10,000,000,000** individuals of **100 kg** each with a power usage of about **10 W/kg**, has a total power of $\mathbf{10^{13}}$ **Watt** which, on the Kardashev scale, equates to a $\boldsymbol{K}_{\mathbf{humanity}}$ value of:

$$\boldsymbol{K}_{umanity} = \frac{log_{10}\left(\mathbf{10^{13}}\right) - 5}{11} \cong \mathbf{0.7}$$

Therefore, given our decidedly early technological level, an animal species must have a Kardashev level of at least **K** = **0.7** to be recognized as **intelligent** by other civilizations. And it is this value of **K** = **0.7** that we will need to keep in mind when we later talk about ETCs. Given the characteristics indicated above, the base level for intelligence that we will refer to is precisely that reached by our species **K** = **0.7**.

However, in the following pages, we will no longer be able to follow a general predictive model in which it is possible to obtain each stage from the previous one, as in the already seen cases of the emergence of animals or prokaryotic and eukaryotic cells. Now we will illustrate the main stages, in the order in which they occurred, that led to the evolution of man. This is because the information we have on this last process is reduced to the single case of *Homo sapiens*. This is therefore an important difference, which must be taken into account for the evaluation of the entire process described in this volume.

We also remember that the various phases that have characterized the evolution of terrestrial life have been "punctuated" by mass extinctions, episodes in which there was a drastic and rapid (in geological terms) decrease in the biodiversity of our planet. Such extinctions have occurred at more or less regular intervals: in the last **250 million** years, the cycles of extinctions—naturally, of different intensity—would have occurred with a regularity of about **26 million** years [90]. Finally, we want to emphasize that extinctions do not have a purely negative value. By eliminating or reducing certain taxonomic groups, they allow others, which remained in the shadow of the dominant organisms, to have a chance and bring new evolutionary solutions. If there had not been the end-Triassic extinction (around **201 million** years ago), dinosaurs would not have become the dominant forms of life on emerged lands, nor today could we enjoy birds. Without the Cretaceous/Tertiary crisis (K/T crisis), which ended the Mesozoic (around **66 million** years ago), mammals would have remained in the shadow of dinosaurs and would not have reached their current sizes... and we would not exist!

15.1 The Birth of Intelligence, Phase by Phase. The Starting Point; The Ediacara Fauna

The term "Ediacara fauna" refers to those associations of organisms found in sedimentary deposits between **575 million** years and **541 million** years ago (the last period of the Neoproterozoic) (Fig. 15.1).

These faunas are composed of unusual animals, on whose lifestyle much ink has been, and still is, spilled [74]. However, alongside these, various traces

Fig. 15.1 The Ediacara fauna (**620–541 My**) (**Ryan Somma**)

in the sediments suggest the existence of worms or other bilaterally symmetrical organisms (animals with bilateral symmetry and an antero-posterior axis of polarity) or their ancestors. It is in this large group of metazoans that intelligent life forms evolve according to the definitions given in the previous section.

15.2 The First Phase; The Increase in the Size of Metazoans and the Acquisition of the Nervous and Vascular Systems

Although some exceptions can be found, one of the main laws found in the animal world is that of the increase in body mass over time. The Ediacara fauna is followed by the Cambrian explosion of life (geological period between **541–485 million** years ago), a phenomenon that, in reality, lasts a few tens of millions of years and continues into the following period, the Ordovician (**485–444 million** years ago).

Practically, all current phyla appear (the different anatomical plans on which metazoans are built), plus others now extinct. Mineral tissues begin to spread in the animal kingdom and the first predators evolve [39]. Not only that, but during the Cambrian, the size of animals begins to become important: we can encounter some that exceed half a meter in length. From the lower Ordovician, we even know of fossil cephalopods whose shell exceeded 1 m in size (Fig. 15.2).

What produces this explosion of organisms, equipped with complex anatomical systems and organs, many of which, but not all, also have calcareous exoskeletons that facilitate their preservation in sediments? Various solutions have been proposed: from the appearance of eyes to other causes, inherent in animal physiology or related to environmental changes. Certainly, the cause of all these changes was not unique; various factors must have intervened. Among these, undoubtedly, some suitable geochemical conditions were produced in the oceans. Among other things, an ever-increasing supply of chemical elements, due to the alteration of terrestrial rocks, was channeled towards the seas, favoring the development of plant and animal life in them.

The achievements indicated in this section (increase in size, acquisition of nervous and vascular systems), occur in different animal phyla at different times, covering a few million years in each phylogenetic line: the probability

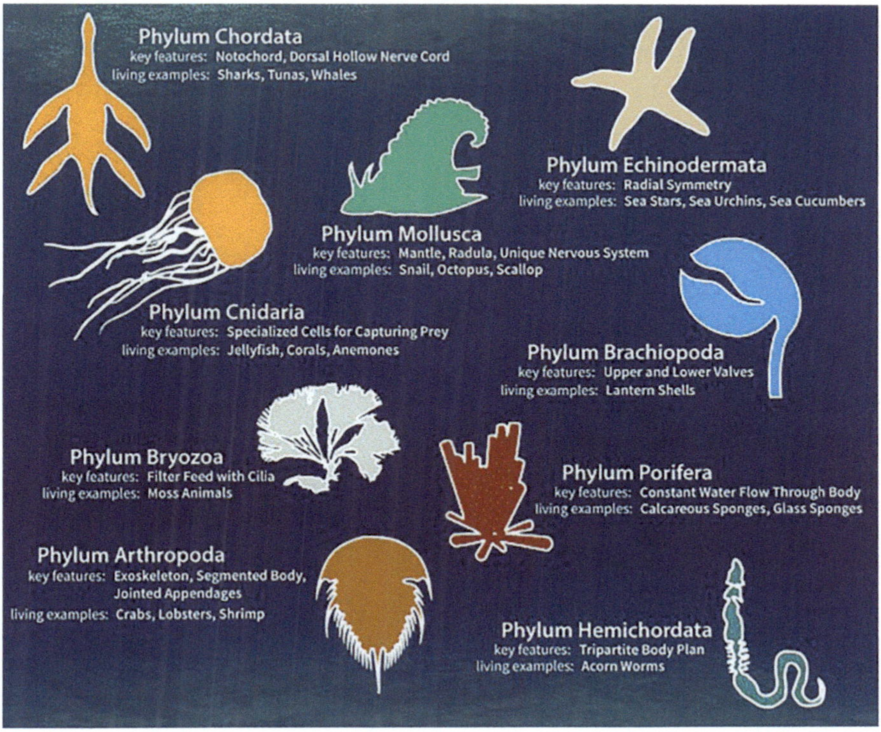

Fig. 15.2 The Cambrian explosion **(541–485 My) (Eric Cheng/STANFOD UNIVERSITY)**

is estimated between 0.02 and 0.04 every 500,000 years with a mass extinction limit of 10,000,000 years.

PHASE 1

a_1	b_1	ΔT_1	ΔT_{01}
0.02	0.04	500,000	10,000,000

15.3 The Second Phase; The Development of Limbs

In many different phyla, once certain sizes and, above all, a certain complexity have been reached, different organisms have developed "limbs", i.e. mobile appendages capable of performing various functions. Within the group of arthropods (whose name means "jointed legs"), which includes insects, arachnids, crustaceans and myriapods, not only do the vast majority of taxa have segmented limbs for moving and performing other activities, but many, particularly insects, are equipped with mandibular elements capable of articulating with each other [60], which make them suitable for performing the most diverse functions and explain the great plasticity of the group (not to mention their extreme abundance, in terms of species) (Fig. 15.3).

Fig. 15.3 Artistic reconstruction of Yohoia; it is an animal from the Cambrian period **(541–485 My)** that has been placed among the arachnomorphs, a group of arthropods that includes the chelicerates and the trilobites **(Junnn11)**

Among the **chordates**, the vast majority of vertebrates develop real symmetrical appendages like the ray fins of fish or the articulated limbs, increasingly complex and developed in the distal sector (the one furthest from the body), in the tetrapod group (vertebrates whose limbs are equipped with a joint with the corresponding bone belt—pelvic or scapular—) and in their ancestors. In tetrapods, indeed, fingers would have evolved not from the rays of the fins of actinopterygian fish, but would turn out to be true and proper evolutionary innovations.

It is believed that the oldest ancestors of the tetrapods (organisms still equipped with fins, although reinforced with an internal axial skeleton), made their appearance already towards the beginning of the Devonian [117], around **410 million** years ago, while the first symmetrical limbs appeared, in vertebrates, probably already from the lower Silurian, more than **430 million** years ago [48].

Finally, it should not be forgotten that, within the phylum of mollusks, the cephalopods developed from the front part of the "foot" the tentacles, modified and prehensile lobes. These are true and proper "arms" capable of performing even very complicated functions. The tentacles appeared with the first cephalopods, in the final part of the Cambrian, more than **500 million** years ago.

As can be seen from the examples reported above, different groups of metazoans develop limbs mainly for locomotion but that, subsequently, can further evolve to adapt to the needs of the organisms that possess them. As in the previous phase, the various phyla have developed their limbs at different times, however, within each group, they always do so after having acquired a certain complexity and convenient nervous and vascular systems: the probability is estimated between 0.01 and 0.02 every 500,000 years with a mass extinction limit of 10,000,000 years.

PHASE 2

a_2	b_2	ΔT_1	ΔT_{02}
0.01	0.02	500,000	10,000,000

15.4 The Third Phase; The Conquest of Land

Without wanting to belittle the intelligence of cetaceans and cephalopods, the only organisms that have developed a technological level like the one we are looking for have evolved on land. We do not know if, in an aquatic environment, it would have been possible to obtain civilizations comparable to the current ones, but we continue to follow the thread that leads us towards our species, as previously announced. The next step, therefore, is the one that makes us gain the land. Indeed, the two previous phases were completely carried out at sea.

The testimonies of terrestrial activities that are recorded during the first phase of the Paleozoic Era (**541–252 million** years ago) are rare if not unique. These are traces of arthropods that have moved on land, it is not known whether to go from one puddle of water to another, or for another reason. The cause that prevented animals from settling stably on the emerged lands does not seem to have been, as once believed, the lack of a suitable atmosphere capable of protecting organisms from ultraviolet rays. We saw in the previous section that, after the NOE, the level of oxygen in the atmosphere was more or less of the same order of magnitude as the current one. Now, it would not have been difficult for an arthropod, protected by an exoskeleton, to walk on land in the sunlight. In reality, the problem seems to have been another: on land the environments were practically sterile, devoid of vegetation. After all, it is always the plants that arrive first to occupy a new environment, creating the conditions for the colonization of animals.

Although traces of spores attributed to terrestrial plants are already found in the middle Ordovician, since about **475 million** years ago, the first vegetable fossil remains of a certain size are found only much later, in the Silurian (**444–416 million** years ago). In an extraordinary conservation deposit discovered near the village of Rhynie [34], in Scotland, and dated to **410 million** years ago, there are clues that make us understand why the conquest of the emerged lands does not go back directly to the time of the NOE (Fig. 15.4).

Fig. 15.4 Reconstruction of the flora of Rhynie (Scotland), in the lower Devonian (**about 410 My**)

In fact, among the exceptional remains of Rhynie, plant roots have been found that present symbiotic fungi, as is the case in the vast majority of current terrestrial plants [111].

Therefore, what determined the time of colonization of plants were not the conditions of the atmosphere, but those required for the establishment of a plant-fungus symbiosis (the symbiosis constitutes an essential process for the evolution of living beings, as we have been able to ascertain during the emergence of the eukaryotic cell). This is indeed necessary to allow plants not only to develop a support root system but above all to allow them to obtain (mainly thanks to the fungal hyphae) water and mineral elements from the soil. That is, it was necessary to wait for the plants to meet the "good" fungi to create the symbiosis capable of making them fit to live on the emerged land.

Once the plants settled on the continents, the metazoans also settled stably in turn. At Rhynie, fossil remains of various terrestrial arthropods have been found. In any case, by the end of the Devonian period (**419–359** million years ago) the plants were well installed on the continents: many stretches of coast were occupied by forests made up of real "trees", even **30 m** high.

It is in this context that the first tetrapods appear, even if their limbs did not yet have the ability to support the animal during its movements out of the water. Rather they served as "paddles" specialized for movements in the liquid element, among the rich aquatic vegetation of coastal environments. However, even before the end of the Devonian, some traces testify to the existence of tetrapods capable of moving, at least temporarily, on land.

Between the Carboniferous (**359–299** million years ago) and the Permian (**299–252** million years ago) mollusks conquer the continents [107], although, at least initially, they probably remained confined to humid environments. Note, however, that despite the fact that, currently, terrestrial mollusk species (freshwater or land) are more numerous than those that live in the seas and oceans, cephalopods (which group the mollusks endowed with intelligence) constitute a set that has always been EXCLUSIVELY marine. Having said that, strictly intelligent terrestrial animals are limited to vertebrates, so we will limit ourselves, in the following stages, to following the evolution of tetrapods.

Although in this case too the different phyla conquer the continents independently, the colonization of each animal group must necessarily wait for that of the plants. Before the end of the Paleozoic Era, various phyla of metazoans have now settled on land: the probability is estimated between 0.02 and 0.05 every 500,000 years with a mass extinction limit time of 10,000,000 years.

PHASE 3

a_3	b_3	ΔT_3	ΔT_{03}
0.02	0.05	500,000	10,000,000

15.5 The Fourth Phase; The Differentiation of Terrestrial Animals

To exploit the resources of the continents, it is not necessary to completely abandon the liquid environment. In fact, most modern amphibians depend on the proximity of ponds and puddles for reproduction. Additionally, their skin requires a certain level of humidity to survive: few species have been able to colonize arid environments (although some Australian frogs have succeeded by employing measures to retain water, living buried and exploiting rare seasonal rainfalls for reproduction).

Although the ancient terrestrial tetrapods differentiated morphologically and taxonomically [96], the great diversification of vertebrates on the continents only occurred when such organisms definitively broke away from the liquid environment for reproduction. Thus, we must wait for the emergence of **amniotes**, a group that includes all reptiles, birds, and modern and fossil mammals. What sets amniotes apart from other tetrapods?

The key to their success is the amniotic egg (Fig. 15.5), a true evolutionary novelty which enabled them to free themselves from the liquid environment for reproduction. Rather, they transferred this environment inside the egg itself. The embryo is immersed in the amniotic fluid, enclosed by the **amnion** which, being impermeable, recreates the aquatic environment necessary for development inside it. The **chorion**, the outermost layer, is permeable to gases, allowing the embryo to breathe and exchange gases, aided by the **allantois**.

Since it is very difficult to recognize an amniotic egg as a fossil, the identification is made from bone remains. The oldest known amniotes are *Hylonomys lyelli* and *Protoclepsydrops haplous*, both found in Canadian Carboniferous sediments about **310 million** years old [112]. Starting from these "precursors", amniotes became capable of colonizing all terrestrial environments, even the most arid. They could thus move away from coastal and marshy areas to conquer even the most remote locations of the continents. Amniotes, therefore, diversified quickly: from the base of the Permian period, we find ecosystems with "reptiles" occupying various ecological niches: herbivores, carnivores and omnivores. Their dimensions increased further (but this also

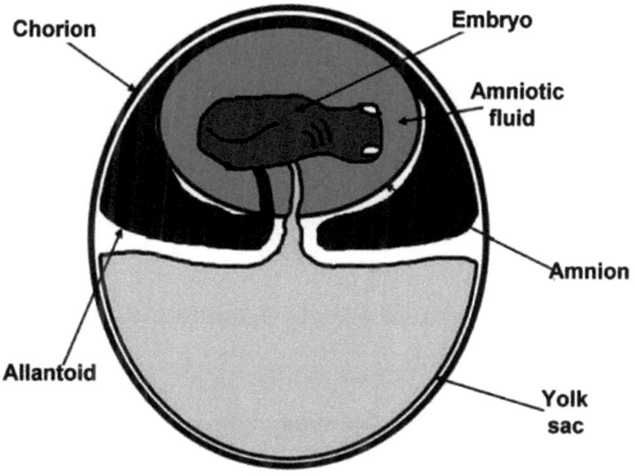

Fig. 15.5 Scheme of the amniotic egg: the embryo bathes in the amniotic fluid, delimited by the waterproof amnion; the yolk sac contains the nutrients for the embryo; the allantois is the embryonic appendage that contains waste and plays a role in respiration; the chorion is the outermost layer of the egg (Carboniferous **310 My**) **(IJA 14/06/2023 Mieli, Valli, Maccone)**

applied to amphibian tetrapods that remained tied to the liquid environment).

Starting from the appearance of the amniotic egg, it took a few million years to witness the differentiation of terrestrial vertebrates and the occupation of niches present in the new environments: the probability is estimated between 0.5 and 1 every 500,000 years with a time limit of microcatastrophe of 10,000,000 years.

PHASE 4

a_4	b_4	ΔT_4	ΔT_{04}
0.5	1	500,000	10,000,000

15.6 The Fifth Phase; The Acquisition of Sociality

We now introduce the fifth phase: **the acquisition of sociality**. This is a further level compared to the first 3 already seen with Macrointervals **A, B** and **C**; the *prokaryotic* cell, the *eukaryotic* cell obtained from the symbiosis of various prokaryotes, the *multicellular beings* composed of various eukaryotic cells all possessing the same genetic code.

The new level is obtained by bringing together various elements of the previous level, that is different individuals of the same species, who come together to operate in a homogeneous way and increase their reproduction, food search and security [1]. In other words, enhancing their own hope of life. Within the group of metazoans, four different degrees of sociality can be recognized: **(A)** *colonial invertebrates*, **(B)** *social insects*, **(C)** *societies of mammals and birds*, **(D)** *human societies*.

The first case **(A)**, includes animals that, like corals, form colonies of hundreds or thousands of individuals capable of communicating via tactile signals (as they are physically connected). Social insects **(B)**, like termites, ants or social bees, however, generally present a caste system (soldiers, workers, reproducing individuals) formed by morphologically distinct individuals. Although tactile signals persist, communication is mainly carried out via chemical substances that various subjects can exchange or leave on the ground for others. The societies of mammals and birds **(C)**, are characterized by forming troops comprising various individuals during reproduction or feeding, aiming to increase the safety of the entire group.

Many of these (most primates, for example) practice varying degrees of parental care, which helps strengthen social bonds between members of different generations. Communication occurs mainly via chemical, visual and auditory signals (however, among monkeys, the habit of *grooming* and mutual cleaning helps reinforce cohesion and social status among individuals). Finally, the last group includes human societies **(D)**, of which any of us is part (the asocial human, the hermit, is a rarity, the exception proving the rule!). Individuals communicate via articulated language (which we will also discuss later, in the last phase) which has allowed, among other things, the evolution of numerous languages and dialects.

As can be inferred from the cases presented, the difference between the various degrees of sociality consists mainly in the different types of signals that individuals of the group are able to exchange to communicate. But if the simplest type of sociality is formed by colonial invertebrates, which are actually very ancient, why discuss the topic only now? The reason is that the various degrees of sociality do not represent a ladder where higher levels derive from lower ones. This is true only for the last two, **C** and **D**, but not for the first two. We can be confident that human societies evolved from those of primates, from which we descend. For this reason, sociality is introduced only now, after the appearance of the ancestors of mammals and birds (Fig. 15.6).

Sociality is quite common among certain mammalian groups, including primates. So at the time of their appearance (between the late Cretaceous and early Eocene, between **70** and **55 million** years ago) such a feature must have

Fig. 15.6 An artistic reconstruction of a social group of *Filikomys primaevus* in a burrow (Cretaceous **75.5 My**) **(Misaki Ouchida/Gregory P. Wilson Mantilla/University of Washington)**

developed quite rapidly: the probability is estimated between 0.4 and 0.8 every 500,000 years with a microcatastrophe time limit of 10,000,000 years.

PHASE 5

a_5	b_5	ΔT_5	ΔT_{05}
0.4	0.8	500,000	10,000,000

15.7 The Sixth Phase; The Upright Stance and Manual Skills

The next stage, in reality, consists of two phases that can occur independently: the acquisition of upright stance and that of manual dexterity. Only in our species do they coexist. The acquisition of upright stance, among modern mammals, is typical of humans but also kangaroos. However, among current primates, we are the only ones to rely on the bipedal position for locomotion (although gibbons can walk upright while balancing with open arms [30]). However, it seems that in the past, at least one other primate resorted to this type of gait, even if thought to be more similar to that of gibbons than our own [80, 108]. Despite this exception, bipedalism is the characteristic that allows placing a fossil hominid in the restricted group from which our species evolved (Fig. 15.7).

In any case, even before mammals achieved upright stance, this condition had been reached by dinosaurs [9]. In fact, this peculiarity helps characterize this group of reptiles and was inherited by their avian descendants. All birds, regardless of intelligence level, have a bipedal stature.

It is interesting to note that the American paleontologist Dale A. Russell [93] speculated on a possible descendant of dinosaurs (Fig. 15.8), if they had not gone extinct during the K/T crisis (let's remember that without dinosaur extinction, mammals would have remained confined to their niche). The scientist considered that a representative of the Troodontidae (=Stenonychosauridae) family could have evolved into a "dinosauroid", an intelligent humanoid being derived from the great Mesozoic reptiles.

Naturally, its hypothesized form, which appears similar to ours, was conceived from the anthropomorphism typical of our species. This does not detract from the fact that an American paleontologist proposed the evolution of a life form with intelligence comparable to ours, starting from a different tetrapod group. What was special about this dinosaur family compared to others, to be chosen as the dinosauroid's origin? Two features, mainly: a relatively high encephalization quotient compared to contemporaries and

Fig. 15.7 Pair of *Ardipithecus ramidus* (**4.4 My**, Pliocene)

forelimbs whose fingers had a certain degree of opposability. We have therefore identified the two abilities that, nowadays combined, result as our characteristic prerogative: upright stance and manual dexterity.

Manual dexterity is a typical ability of animals that need to grasp objects, like arboreal mammals, the most characteristic being primates. These are equipped with adaptations for arboreal living (stereoscopic vision, opposable thumb, etc.). The oldest known representative, from a partially preserved skeleton, has been dated to about **55 million** years ago, just a decade after the dinosaur extinction.

The opposable thumb is a primate hallmark, but no monkey can touch the pad of the same hand's fingers with its thumb. Although some australopithecines (the "robust" type belonging to *Paranthropus*) are thought to have had hand anatomy capable of producing simple stone tools [73], human manual dexterity has evolved well beyond the possibilities of primitive hominids and ancestors.

Fig. 15.8 Stenonychosaurus (**76 My**) and Dale Russell's study for the "dinosauroid" (**sculptures by Ron Seguin**)

This phase brought us to the threshold of our genus' appearance, a leap of several tens of millions of years compared to the previous phase. But this is justified by the occurrence of two characters that, although they can develop independently, must work together (and reach a relatively sophisticated evolutionary level) to drive evolution towards human-like intelligence: the probability is estimated between 0.005 and 0.01 every 500,000 years with a microcatastrophe time limit of 10,000,000 years.

PHASE 6

a_6	b_6	ΔT_6	ΔT_{06}
0.005	0.01	500,000	10,000,000

15.8 The Seventh Phase; The Change in Diet and the Growth of the Brain

The brain requires a high energy cost as well as a significant protein contribution for its constitution. Its development and increase have required a strong supply of valuable resources to allow adequate growth.

For this reason, the anthropologist Craig B. Stanford in 2001 [105] hypothesized that a dietary shift from the herbivore/omnivore diet typical of australopithecines to a more carnivorous one was important for human evolution. The abundance of animal carcasses killed by numerous African Plio/Pleistocene carnivores, combined with hominids' gregarious habits (which would have allowed them to successfully contest prey from mammalian predators), may have allowed a population access to a more carnivorous diet (Fig. 15.9).

The significant protein intake obtained could have favored an increase in brain size compared to other primate populations. The proposal sparked various reactions, but beyond the hypothesis' correctness, if we compare the first representatives of our genus with their australopithecine precursors,

Fig. 15.9 Male of *Homo erectus* **(2–0.6 My)** Powered by OpenAI

important morphological changes are noticeable, such as a relative reduction of the masticatory apparatus (a *shift* from a more fibrous diet to one poorer in these foods) as well as an increase in brain volume.

Data in hand, even without hypothesizing our ancestors' exact diet, it is possible to find a correlation between a dietary change and the brain increase that occurred toward the beginning of the Pleistocene, between **2.5** and **1.5 million** years ago. The timeframe in which it is realized is relatively short, less than **1 million** years: the probability is estimated between 0.05 and 0.1 every 500,000 years with a microcatastrophe time limit of 10,000,000 years.

PHASE 7

a_7	b_7	ΔT_7	ΔT_{07}
0.05	0.1	500,000	10,000,000

15.9 The Eighth Phase; The Organization of the Brain on Abstract Thought

The next phase involves the ability to conceive abstract thoughts. Although a certain number of animals possess some capacity for abstraction, for example, the ability to handle numbers and different quantities of objects, human possibilities go far beyond. The abilities related to cognitive functions, therefore to calculation, abstract thought and also to language, are linked to the development of the frontal lobes (Fig. 15.10) and the convolutions of these brain regions [41].

If we evaluate the evolution of the human brain from our australopithecine ancestors using appropriate endocranial casts, we observe that starting from *Homo erectus*, the organ significantly developed in the aforementioned regions. In particular, in modern humans, brain evolution is manifested mainly by an enlargement at the coronal cranial suture.

The increase in endocranial capacity was achieved differently by Neanderthal man, *Homo neanderthalensis*, and modern man, *Homo sapiens*. If we compare the skulls of these two species, we notice that the Sapiens' is more "rounded", taller than the other, which is more elongated antero-posteriorly.

Apart from brain proportions, what would be the first manifestations of human abstract thought? It is difficult to answer this question. However, it is possible they already manifested with *H. erectus*. A particular artifact found on the island of Java is indeed attributed to this species: a mollusk shell decorated with a "zigzag" pattern, made without any apparent practical reason [110].

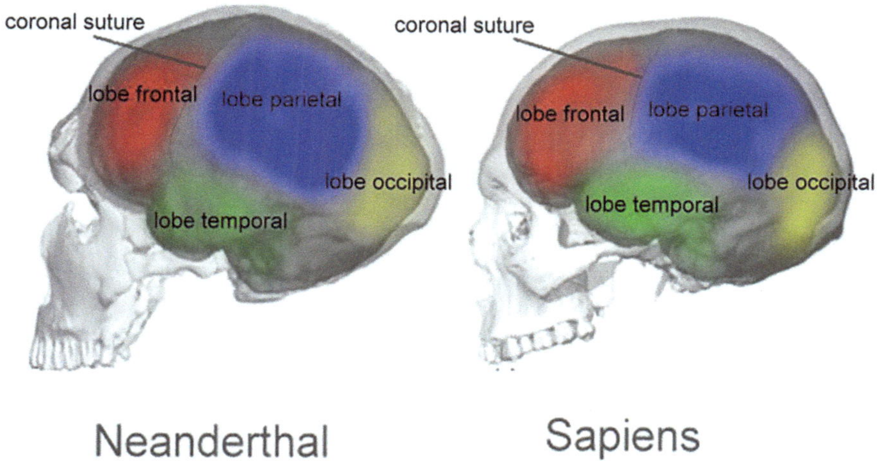

Fig. 15.10 Simplified diagram showing the human brain, with the various brain lobes indicated **(IJA 14/06/2023 Mieli, Valli, Maccone)**

No symbolic activity has yet been associated with Neanderthal man, although at the brain level, all the conditions for realizing such conceptions seem to be met. Perhaps it was just a matter of time...

Despite current abstract abilities far exceeding the decorated shell illustrated previously, it is quite possible that starting from the emergence of *H. erectus*, certain human populations acquired the ability of abstract thought. In this case, a few hundred thousand years would have been enough for its establishment: the probability is estimated between 0.5 and 0.9 every 500,000 years with a microcatastrophe time limit of 10,000,000 years.

PHASE 8

a_8	b_8	ΔT_8	ΔT_{08}
0.5	0.9	500,000	10,000,000

15.10 The Ninth Phase; The Birth of Articulated Language and Technique

One of the characteristics that distinguishes us from all other animals is undoubtedly articulated language. Although great apes are capable of learning rudimentary sign languages to communicate with their tutors, they are unable, due to their anatomy, to express themselves with the sound range and articulations that we have.

The morphological characteristics that allow us such performance are linked to the morphology of the larynx and the hyoid bone [12], located at the base of the tongue, between the jaw and the thyroid cartilage of the larynx itself (Fig. 15.11).

It is not easy to accurately reconstruct the anatomy of the vocal region from disarticulated bone remains. However, the hyoid bone of a fossil hominid can be found, studied, and compared with ours. In particular, one belonging to *H. erectus* has been discovered.

Its morphology is very reminiscent of a modern hyoid bone, although some minor differences suggest a more restricted modulation of the vocal tract that probably limited the use of articulated language. A Neanderthal hyoid bone is also known [5], even more similar to that of modern humans than *H. erectus*. The fossil evidence suggests, therefore, that the ability to emit complex vocalizations existed at least from the common ancestor between Neanderthals and modern humans, even if this does not mean they could already communicate like us.

Regarding technology, we know *H. neanderthalensis* was capable of a refined lithic industry, not inferior to that of *H. sapiens* of the time, although the latter also expressed themselves at other levels, including cave painting (but we do not know if all known cave paintings were made by our species!). Naturally, we were still far from modern technical capabilities.

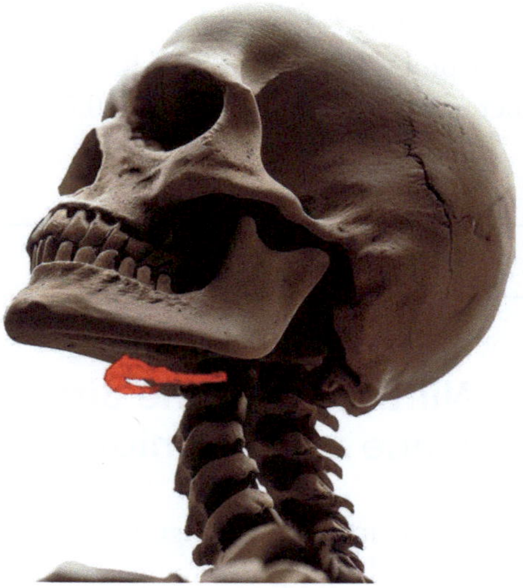

Fig. 15.11 Morphology and position of the hyoid bone (in red) Powered by OpenAI

However, articulated language was necessary to explain and transmit (first orally and then in writing) the technical instructions for producing increasingly complicated objects and, therefore, to advance technology from prehistory to current levels. It was also necessary to be able to express abstract thought: the probability is estimated between 0.4 and 0.8 every 500,000 years with a microcatastrophe time limit of 10,000,000 years.

PHASE 9

a_9	b_9	ΔT_9	ΔT_{09}
0.4	0.8	500,000	10,000,000

Thus, at the end of these last **nine** stages, we have arrived at the emergence of *H. sapiens*, the only species that will show itself endowed with the intelligence described at the end of the initial section of Macrointerval **C**. Thanks to its characteristics, our species has acquired a technical capacity beyond all prediction; to the point of being able to send messages capable of reaching solar systems far away from ours. Naturally, all this did not happen overnight; it took a couple of hundred thousand years from its origin.

15.11 Evaluation of Probabilities at Each Stage

We have thus obtained the **36** input values to insert in STEP 1 of the algorithm for calculating the lognormal statistical distribution of Maccone (Table 15.1). The Fig. 15.12 shows the lognormal distribution related to macrointerval **C**.

Table 15.1 Fifth Drake—macrointerval **C**: The **36** values of the frequencies a_j and b_j minimum and maximum, of the time ΔT_j of observation and the time ΔT_{0j} of microcatastrophe, for each phase described in the previous paragraph.

Phase	Description	a_j	b_j	ΔT_j	ΔT_{0j}
1	Increase in metazoan size	0.02	0.04	500,000	10,000,000
2	Limb development	0.01	0.02	500,000	10,000,000
3	Conquest of land	0.02	0.05	500,000	10,000,000
4	Differentiation of terrestrial animals	0.50	1.00	500,000	10,000,000
5	Acquisition of sociality	0.40	0.80	500,000	10,000,000
6	Upright stance and manual dexterity	0.005	0.01	500,000	10,000,000
7	Diet change and brain growth	0.05	0.10	500,000	10,000,000
8	Organization of abstract thought	0.50	0.90	500,000	10,000,000
9	Birth of articulated language and technique	0.40	0.80	500,000	10,000,000

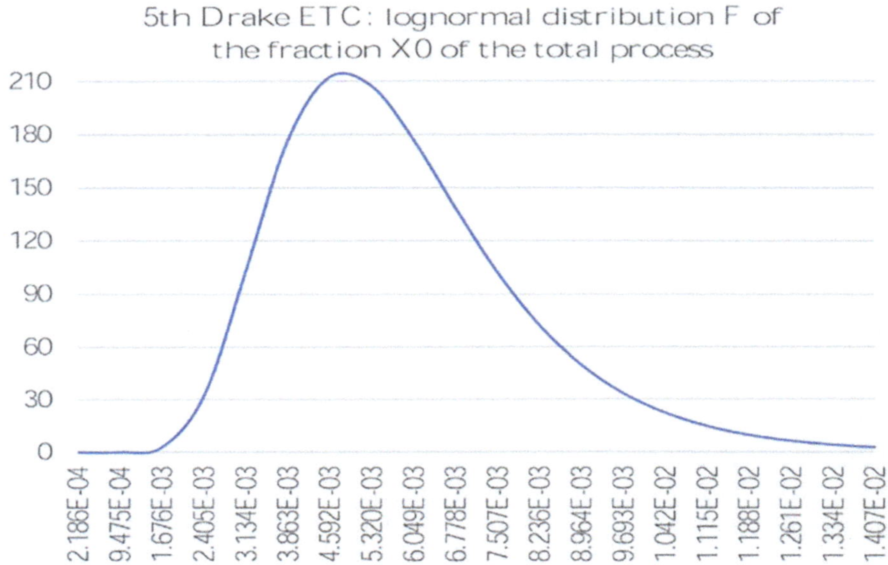

Fig. 15.12 Fifth Drake—macrointerval **C**: A lognormal distribution **Φ** of the process in the medium period ΔT0: the average value is **5.89 · 10-3**, the standard deviation is **2.19 · 10-3 (IJA 14/06/2023 Mieli, Valli, Maccone)**

Reporting, with the method of the three previous cases, the probability of ETC $\mathbf{2.19 \cdot 10^{-3}}$ of the medium period $\mathbf{\Delta T_0}$ equal to **90 My** over the long period $\mathbf{\Delta T}$ equal to **500 My**, we obtain the probability of macrointerval B:

$$\mathbf{f_C = 3.43\%}$$

included between the two minimum and maximum values

$$\mathbf{f_{C\,min} = 1.25\%} \text{ and } \mathbf{f_{C\,max} = 5.66\%}.$$

As you can see, unlike the two macrointervals **A** and **B**, in this last one the probability is decidedly more contained.

16

Evaluation of the Total Probability: The Fifth Drake Parameter

Having defined the probabilities of the three macrointervals **A**, **B** and **C** necessary to describe with more accuracy the probability of intelligent life starting from bacteria, we now combine them into a new total lognormal that will represent the fifth Drake parameter.

We must note that applying Maccone's lognormal method to only three parameters cannot be rigorous because it does not respect the conditions of the central limit theorem which ensures that a sufficiently high number of random variables converge towards a *normal* (in our case *lognormal*) distribution; however, since in this context we are only looking for the order of magnitude of the Drake parameters, we believe this method is fully justified.

We have, in conclusion, obtained **six** input values to be inserted in STEP 2 of the algorithm for calculating the statistical distribution lognormal of Maccone (Table 16.1).

We note that for the macro intervals, we skip STEP 1 entirely because we have the final frequencies $\mathbf{A_j}$ and $\mathbf{B_j}$ of STEP 2. Furthermore, the time frame for the realization of the entire process is the sum of the three times of macro catastrophe of the macro intervals **A**, **B** and **C** or:

$$\mathbf{0.5\,Gy \cdot 3 = 1.5\,Gy}$$

Table 16.1 Fifth Drake—**TOTAL**: The six values of the frequencies A_j and B_j minimum and maximum in the total time of **1.5 Gy**

		A_j	B_j
1	Eukaryotes	0.289	0.709
2	Metazoa	0.597	0.944
3	Homo	0.013	0.057

E. Mieli et al., *The Living Galaxy*, https://doi.org/10.1007/978-3-031-67324-5_16

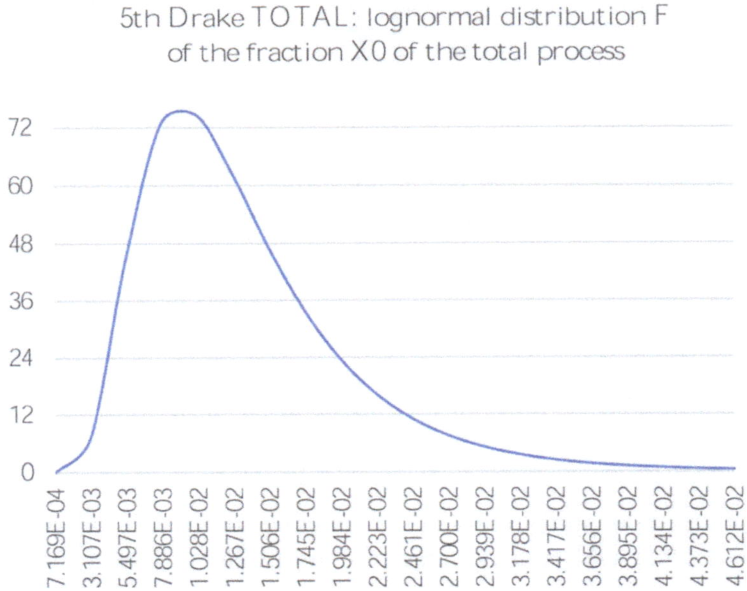

Fig. 16.1 Fifth Drake—**TOTAL**: The lognormal distribution **Φ** of the process in the medium term **ΔT0**: the average value is $\mathbf{1.34 \cdot 10^{-2}}$, the standard deviation is $\mathbf{7.17 \cdot 10^{-3}}$ **(IJA 14/06/2023 Mieli, Valli, Maccone)**

The Fig. 16.1 shows the lognormal distribution of the entire process **A**, **B** and **C** that directly provides the final values of the fifth parameter without having to transform the probabilities from medium to long term. The final value of the fifth parameter is $\mathbf{f_i}$ = **1.34%** included between the minimum and maximum values:

DRAKE 5

$\mathbf{f_{i\,min}}$	$\mathbf{f_{i\,max}}$
$\mathbf{6.0 \cdot 10^{-3}}$	$\mathbf{2.1 \cdot 10^{-2}}$

17

Considerations on the Fifth Parameter

Unlike the fourth Drake parameter, which we calculated around **0.5** (probability of **50%**) over a period of **100 million** years in the previous section, the fifth Drake parameter is just above **0.01** (probability of **1%**) over a period of **1500 million** years (excluding GOE and NOE), therefore significantly lower. This does not surprise us because, after all, *H. sapiens* is the only successful attempt among an infinity of species that have appeared on the planet. Moreover, the subdivision of the calculation into three macro intervals has highlighted two things:

(a) The essential role that oxygen played in determining the timing of evolutionary processes: see both the GOE and the NOE, which with their occurrence, have accelerated the times and given a decisive impulse towards the eukaryotes, first, and towards the metazoans, after;

(b) The bottleneck highlighted precisely in the last step, namely the onset of intelligence whose probability was calculated around **3%** in **0.5 Gy**, while neither the appearance of eukaryotes (probability of **50%** in **0.5 Gy**), nor that of metazoans (probability of **85%** in **0.5 Gy**) have proved particularly problematic from our calculations.

E. Mieli et al., *The Living Galaxy*, https://doi.org/10.1007/978-3-031-67324-5_17

18

The Oxygen Curve

At this point we want to make some considerations on the well-known curve of oxygen growth reported in Fig. 18.1.

We have found, in the calculation of the fourth parameter, a good probability of development of prokaryotes (**50%**) in **100 My,** starting from at least **3.7 Gy**. The positioning of the first photosynthetic prokaryotes is controversial, but in any case certainly the first oxygen produced was reabsorbed by oxidative chemical phenomena.

When these ceased, the oxygen released by prokaryotes began to invade the planet, reaching about **1%** of the current value at **2.1 Gy** (GOE). We found

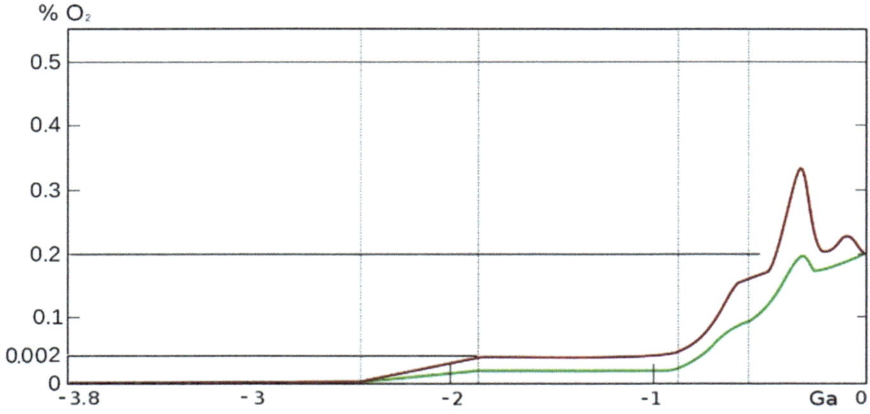

Fig. 18.1 The curve depicted shows the rise in the concentration of oxygen in the atmosphere during the GOE from **0%** to **0.2%** between **2.4** and **1.9 Gy** and, during the NOE, between **0.8** and **0.5 Gy**. The green and red curve represents the minimum and maximum value hypothesized **(IJA 14/06/2023 Mieli, Valli, Maccone)**

E. Mieli et al., *The Living Galaxy*, https://doi.org/10.1007/978-3-031-67324-5_18

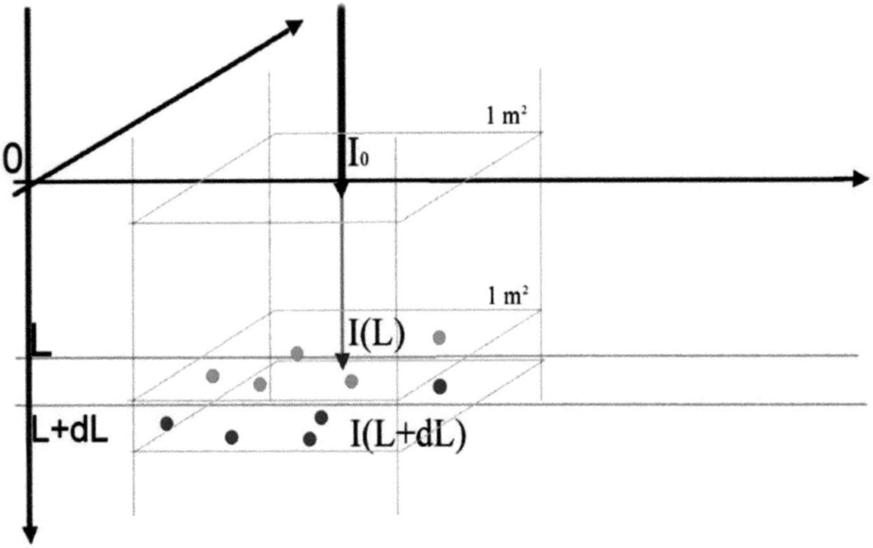

Fig. 18.2 Light penetration **I(L)** between the depth **L** and the depth **L + dL** on a **m2** of sea surface. The small gray discs represent the cells in suspension **(IJA 14/06/2023 Mieli, Valli, Maccone)**

the same probability of about **50%** for the emergence of eukaryotes, but in **500 My** from **2.1 Gy**. Once these also became photosynthetic (around **1.5 Gy**?), the percentage of oxygen began to rise significantly, reaching values slightly lower than the current ones around **0.7 Gy** (NOE). Let's now try to estimate the growth rate of oxygen in these two key events, the GOE and the NOE. Let's consider Fig. 18.2 which describes a marine environment from altitude **0** (sea level) towards the depth **L**. The light intensity **I** (**W/m²**) will be equal to **I(L)** = **I(0)** ≡ $\mathbf{I_0}$ on the sea surface and equal to **I(L)** at the depth **L**.

For each square meter of seabed, between the depth **L** and **L** + **dL**, the number of photosynthetic cells in the volume element will be equal to:

$$\mathbf{N_f = 1m^2 \cdot dL \cdot \rho}$$

where **ρ** is the density, or the number of cells per **m³**. Therefore, let **r** be the linear dimension of the cells and **ω** their absorption coefficient (between **0** and **1**; **0** for totally transparent cells and **1** for totally opaque cells), the effective cross section $\mathbf{S_{eff}}$ (or the one that contributes to the phenomenon) of the cells present in the direction of the light rays on **1 m²**, will be:

$$\mathbf{S_{eff} = 1m^2 \cdot \omega \cdot \rho \cdot r^2 \cdot dL}$$

Consequently, $\boldsymbol{\omega} \cdot \boldsymbol{\rho} \cdot \mathbf{r}^2 \cdot \mathbf{dL}$ is the percentage of *darkening*, due to the **dL** layer, of the light intensity **I(L)** to depth **L**; so the light intensity beyond this layer will be:

$$\mathbf{I}(\mathbf{L}+\mathbf{dL}) = \mathbf{I}(\mathbf{L}) \cdot \left(\mathbf{1} - \left(\omega \cdot \rho \cdot \mathbf{r}^2 \cdot \mathbf{dL}\right)\right)$$
$$\mathbf{I} + \mathbf{dI} = \mathbf{I} - \mathbf{I} \cdot \omega \cdot \rho \cdot \mathbf{r}^2 \cdot \mathbf{dL}$$
$$\mathbf{dI} = -\mathbf{I} \cdot \omega \cdot \rho \cdot \mathbf{r}^2 \cdot \mathbf{dL}$$
$$\mathbf{dI} / \mathbf{I} = -\omega \cdot \rho \cdot \mathbf{r}^2 \cdot \mathbf{dL}$$

Integrating both members between **0** (sea level) and **L**, we have:

$$\left[\mathbf{ln}(\mathbf{I}) - \mathbf{ln}(\mathbf{I}_0)\right] = -\ \omega \cdot \rho \cdot \mathbf{r}^2 \cdot \mathbf{L}$$
$$\mathbf{ln}(\mathbf{I} / \mathbf{I}_0) = -\ \omega \cdot \rho \cdot \mathbf{r}^2 \cdot \mathbf{L}$$

and, finally, by extracting the exponential:

$$\mathbf{I} / \mathbf{I}_0 = \mathbf{exp}\left(-\omega\rho\mathbf{r}^2\mathbf{L}\right)$$
$$\mathbf{I}(\mathbf{L}) = \mathbf{I}_0\ \mathbf{exp}\left(-\omega\rho\mathbf{r}^2\mathbf{L}\right)$$

In this way, solving the differential equation for separation of the variables, the light intensity ($\mathbf{W\ m^{-2}}$) is obtained which comes to depth **L** starting from intensity $\mathbf{I_0}$ to sea level:

$$\boldsymbol{I(I_0, L, r, \rho, \omega) = I_0 e^{-\omega\rho\tau^2 L}}$$

Therefore, similarly to cells suspended in the liquid, the water molecules themselves also reduce the intensity according to a similar law; thus the complete equation of the reduction of light intensity caused by the water and the cells in suspension is written by adding, to the exponent, the factor due to the absorption of water molecules:

$$\boldsymbol{I\left(I_0, L, r, \rho, \omega, \varphi\right) = I_0 e^{-\left(\omega\rho\tau^2 + \varphi\right)L}}$$

We now estimate the world marine surface **S** interested in the phenomenon of photosynthesis: it is the coastal area of the planet whose waters do not exceed **20 m** depth; that is:

$$\mathbf{S} = \mathbf{C} \cdot \mathbf{D} \cdot \zeta = \mathbf{10^9\ m^2}$$

Having place:

C = 10^8 **m** extension of the continental coasts
D = 10^2 **m** Thickness of the continental coast, where the depth does not exceed **20** m
ζ = 10^{-1} fraction of the coasts not directly above the seabed

The **dW** light power, captured by photosynthetic cells with depth **L** on the entire surface **S**, will instead be:

$$dW = S \cdot dL \cdot \rho \cdot \alpha \cdot r^2 \cdot I_0 \cdot e^{-\left(\omega\rho\tau^2 + \varphi\right)L}$$

where α is the fraction of the cell surface interested in photosynthetic absorption. We observe that α must necessarily be less than the absorption coefficient ω which measures all luminous absorption, both photosynthetic and not.

If we add the contribution of all the layers of the seabed, we obtain the total power absorbed by the cells. To obtain this sum, we ideally integrate the latest formula obtained for **dW** between **0** and **+∞** compared to depth **L** (obviously the light absorption is practically zero beyond a certain depth). It is obtained for total power **W**:

$$W(\rho) = \frac{\left(S \bullet \alpha \bullet r^2 \bullet I_0\right) \bullet \rho}{\left(\omega \bullet r^2\right) \bullet \rho + \varphi}$$

If we now make ρ tend to infinity, we obtain the **value of saturation** of the power captured for photosynthesis (turbid waters):

$$W_s = \frac{S \bullet \alpha \bullet I_0}{\omega}$$

This is obviously valid for $\omega \neq 0$, which is the limit of totally transparent cells (moreover, if ω goes to **0**, α, that is less than ω, must do the same, but their quotient does not). As we should have expected, this value is independent of the size of the cells that carry out photosynthesis (saturation value), and of the water solution coefficient φ.

To get I_0 (effective photosynthetic power per m^2), we must multiply the three factors:

$\mathbf{W_{bs}} = \mathbf{10^3}$ (W/m^2) **s**olar **b**ase power to the ground
$\eta = \mathbf{0.154}$ Average sun efficiency in the year
$\gamma = \mathbf{0.3}$ Visible light fraction used

Getting:

$I_0 = \mathbf{46.2}$ (W/m^2) Photosynthetic power for **m**^{2} effective

Therefore:

$$W_s = \frac{\alpha}{\omega} \bullet W_{bf}$$

Having place:

$W_{bf} \equiv \mathbf{S} \cdot \mathbf{I_0} = \mathbf{4.62} \cdot \mathbf{10^{10}}$ W **B**asic **p**hotosynthesis power

To obtain the energy necessary to produce an $\mathbf{O_2}$ molecule, we must multiply the following factors:

$\nu_f = \mathbf{5.45} \cdot \mathbf{10^{14}}$ (s^{-1}) Medium light frequency in photosynthesis
$h = \mathbf{6.63} \cdot \mathbf{10^{-34}}$ (J · s) Plank constant
$\mathbf{n} = \mathbf{10}$ Number of photons for each $\mathbf{O_2}$ molecule

Getting:

$\varepsilon = \mathbf{3.62} \cdot \mathbf{10^{-18}}$ (J) Energy for oxygen molecule produced

Let's now give an estimate of the oxygen molecules present after the GOE and after the NOE starting from the data of the current atmosphere.

$\mathbf{S_\oplus} = \mathbf{5.3} \cdot \mathbf{10^{14}}$ (m^2) Earth surface
$\mathbf{M_a} = \mathbf{10^4}$ (kg/m^2) Air mass for **m**^{2}
$\mathbf{p_{GOE}} = \mathbf{0.2\%}$ GOE oxygen percentage
$\mathbf{p_{NOE}} = \mathbf{20\%}$ NOE oxygen percentage
$\mathbf{M_{Oss}} = \mathbf{2} \cdot \mathbf{10^{-26}}$ (kg) Mass oxygen molecule

There will be the following values:

$\mathbf{N_{GOE}} = \mathbf{S_T} \cdot \mathbf{M_a} \cdot \mathbf{p_{GOE}} / \mathbf{M_{Oss}} = \mathbf{5.3} \cdot \mathbf{10^{41}}$ Number of oxygen molecules GOE
$\mathbf{N_{NOE}} = \mathbf{S_T} \cdot \mathbf{M_a} \cdot \mathbf{p_{NOE}} / \mathbf{M_{Oss}} = \mathbf{5.3} \cdot \mathbf{10^{43}}$ Number of oxygen molecules NOE

We just have to try to evaluate the oxygen produced by a world cyanobacteria population and a similar population of photosynthetic eukaryotes. To do this, we must estimate the two **optical-structural parameters** for prokaryotes and eukaryotes, that is α (fraction of the photoactive surface of the cell) and ω (luminous absorption coefficient of the cell). We therefore place the following values (where the indices '***p***' and '***e***' indicate respectively 'prokaryotes' and 'eukaryotes'):

$$\alpha_p = 1.0 \cdot 10^{-2}$$
$$\alpha_e = 4.0 \cdot 10^{-2}$$
$$\omega_p = 5.0 \cdot 10^{-1}$$
$$\omega_e = 1.0 \cdot 10^{-1}$$

where we chose to attribute to the eukaryotes a fraction of photosynthetically active surface four times greater than the prokaryotes, given that the eukaryotes could probably be more efficient; in addition, it was chosen to attribute to the eukaryotes an absorption coefficient one-fifth of the prokaryotes, given that the eukaryotes are without cell wall.

Finally, remembering that the calculated value of the total power (J/s) captured for photosynthesis is:

$$W_s = \frac{\alpha}{\omega} \bullet W_b \; (\mathrm{J/s})$$

and energy, for oxygen molecule produced, is:

$$\varepsilon = 3.62 \cdot 10^{-18} \; (\mathrm{J})$$

The total number of oxygen molecules produced to the second is obtained for prokaryotes and eukaryotes:

$$N_{Oss\ proc} = \frac{\frac{\alpha_p}{\omega_p} \bullet W_b}{\varepsilon} = 2.56 \bullet 10^{26} \left(\frac{\text{molecules}}{\text{second}} \right)$$

$$N_{Oss\ euc} = \frac{\frac{\alpha_e}{\omega_e} \bullet W_b}{\varepsilon} = 5.11 \bullet 10^{27} \left(\frac{\text{molecules}}{\text{second}} \right)$$

Taking into account that in a million years, there are **1 My** = **3.15** × **10**[13] (**s**), the minimum time of formation of the GOE and NOE is respectively:

$$\boldsymbol{T}_{\mathbf{GOE}} = N_{\text{GOE}} / \left(\text{N}_{\text{Oss proc}} \cdot 3.15 \times 10^{13}\right) = \mathbf{66\,My}$$

$$\boldsymbol{T}_{\mathbf{NOE}} = N_{\text{NOE}} / \left(N_{\text{Oss euc}} \cdot 3.15 \times 10^{13}\right) = \mathbf{330\,My}$$

As you can see, even if photosynthetic eukaryotes are more efficient than their corresponding prokaryotes, the quantity of oxygen of the NOE is **100** times higher; therefore, the time of realization of the NOE remains higher than five times. Obviously, we did not consider the phenomena of reabsorption of oxygen (oxidation, breathing of the heterotrophic, etc.) that expands the estimated times.

Part III

Drake's Social Parameters: f_c and f_l

The sixth and seventh parameters of Drake are rightly traditionally defined as *the social parameters* and complete the descriptive picture of a potential galactic civilization. For example: are we sure that an alien civilization wants to communicate with us? (Fig. III.1). Obviously, addressing these topics is inevitably less solid than the previous ones as it relies on plausible hypotheses without any experimental confirmation.

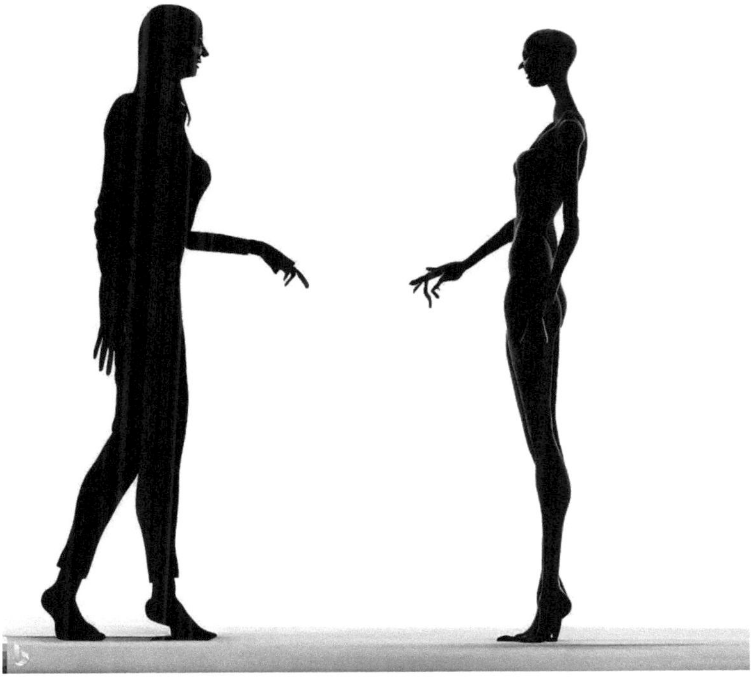

Fig. III.1 Are we sure that aliens want to communicate? Powered by OpenAI

19

Sixth Drake: Fraction of Planets Where Life Decides to Communicate

As mentioned in the introduction, the sixth parameter of Drake, in this work, represents only the fraction of civilizations that freely decide to communicate and not to hide.

While the original sixth parameter intended to count civilizations that become technological and communicative, we have delegated this aspect to the fifth parameter through the definition of intelligence, derived from Kardashev, of a **K** at least equal to **0.7** (ours).

Additionally, we have shifted to the seventh parameter all those cases in which civilizations ARE FORCED or induced not to communicate and, therefore, cease to be visible.

What remains in the sixth parameter is the DELIBERATE social choice of galactic civilizations not to communicate from the beginning of their history. On such an elusive topic, we do not feel able to express scientific opinions and assign a probability of 50% with a deviation of 10%.

DRAKE 6

$f_{c\ min}$	$f_{c\ max}$
$4.0 \cdot 10^{-1}$	$6.0 \cdot 10^{-1}$

E. Mieli et al., *The Living Galaxy*, https://doi.org/10.1007/978-3-031-67324-5_19

20

Seventh Drake: Temporal Fraction of the Duration of a Civilization

With the seventh parameter, we move into a slightly different context from the previous ones because we have to consider not the probability of an event occurring, but the duration of the same event. The lognormal $\mathbf{\Phi(X_0)}$ function of the compound probability $\mathbf{X_0}$ is transformed into another distribution function, $\mathbf{F_p(\Delta T)}$, a function of time $\mathbf{\Delta T}$, the duration of the galactic civilization, and the confidence interval $\mathbf{p}$, which we will define later.

The mathematical problems end here, while the challenges of choosing the input data remain to be faced. It is a matter of hypothesizing which events, completely unrelated to the particular planetary context, can constitute universal bottlenecks that endanger any galactic civilization. We must then also fix the probability of survival and the duration $\mathbf{\Delta T_{0j}}$ of each challenge, beyond which the danger is overcome. This results in the mathematical framework that we will describe in the following sections.

Finally, we reiterate that we have slightly encroached upon Drake's sixth parameter (the percentage of civilizations that choose to communicate), as can be seen from the fourth and seventh challenges (respectively, **spontaneous involution** and the **point Ω**). However, this is not a true misunderstanding because in this context we have mainly analyzed the causes that, in some way, lead civilizations to no longer communicate AFTER A PERIOD OF COMMUNICATION, leaving to the sixth parameter the analysis of the socio-cultural motivations that lead them to voluntarily isolate themselves from the beginning of their history, always hiding from others.

E. Mieli et al., *The Living Galaxy*, https://doi.org/10.1007/978-3-031-67324-5_20

20.1 Is Gott's Delta-T Argument Applicable to the Duration of Galactic Civilizations?

In 1993, astrophysicist Richard Gott published an article titled "Implications of the Copernican principle for our future prospects" [38] in which he attempted to calculate the probable duration of the human race before its inevitable extinction. The genesis of the article began in 1969 when Gott visited Berlin and the infamous Wall. Gott applied mathematical reasoning to try to predict the lifespan of the Wall. He had not visited it in the year of its construction (1961) nor in the year of its demolition (1989), but AT SOME RANDOM POINT DURING ITS EXISTENCE. It was therefore reasonable to assume that his 1969 visit occurred within the middle two quarters of the Wall's lifespan with a **50%** probability. If the visit was at the beginning of the second quarter (only the first quarter had passed), the Wall still had **3/4** of its life ahead, meaning it would stand **three** times longer than the time elapsed since construction. If at the end of the third quarter, the Wall had only **1/3** of its lifespan remaining compared to the time already passed. At that point, the Wall was **8** years old, and Gott concluded there was a **50%** chance the Cold War symbol had between **2.7** and **24** years left. As we know, the Wall was demolished **20** years and a few months after Gott's visit, perfectly within his predicted range. According to Gott, this analysis can predict any temporal event's duration, *provided the observer positions themselves randomly within it.*

We've given the classic **50%** reference probability (**p** = **1/2**) example, but we can extend it to any **p** value. In that case, the ratio between future and past lifespan is expressed by:

$$\frac{\Delta T_{future}}{\Delta T_{past}} = \begin{cases} \dfrac{1+p}{1-p} \; max \\[2ex] \dfrac{1-p}{1+p} \; min \end{cases}$$

In the case of human civilization's duration with **p** = **1/2** and a duration so far of about **200,000** years, a minimum future value of about **70,000** years and a maximum of **600,000** years would be obtained.

But is this approach correct? In our opinion, NO, for two reasons:

Reason A) It violates the main prerequisite of Gott's reasoning—the supposed randomness of the chosen moment for calculation. While Gott's Berlin visit was undoubtedly unrelated to the Wall's history, in a galactic civilization's case, the moment Gott's question is posed can only be immediately after the formulation of abstract mathematical thought, a thousand years more or less. Therefore, this is not any moment of our civilization but the era immediately following the birth of abstract math—the era of Galilei, Einstein, and Gott himself. The delta-T reasoning cannot be applied and, as we will see in the conclusions, it is wrong by excess or defect depending on whether we intend to analyze the entire species' duration or just the technological civilization.

Reason B) not only must the chosen calculation moment be accidental within the process's life, but all the temporal intervals must be equivalent, i.e., the outline conditions (such as human behavior) must remain unchanged. As we will see, this is never true for a complex technological society.

So we will not follow Gott's delta-T reasoning path but instead reformulate Maccone's distribution function applied to the duration $\mathbf{\Delta T}$.

20.2 The Calculation of the Distribution Curve of the Duration of a Galactic Civilization

As in the previous Drake parameters, we proceed to define in order the mathematical magnitudes used:

$\mathbf{\Delta T_{0j}}$ it is the time of the *bottleneck* or *challenge* phase, that is the average time of permanence of the risk conditions of phase $\mathbf{j}$ before its definitive overcoming

$\mathbf{A_j}, \mathbf{B_j}$ these are the minimum and maximum probabilities, of the variable $\mathbf{X_j}$, of survival of the galactic civilization in the period $\mathbf{\Delta T_{0j}}$ of phase $\mathbf{j}$

$$\mu \equiv \sum_j \frac{\mathbf{B_j}\left(\ln \mathbf{B_j} - 1\right) - \mathbf{A_j}\left(\ln \mathbf{A_j} - 1\right)}{\mathbf{B_j} - \mathbf{A_j}}$$

it is the so-called *logarithmic mean* of the lognormal distribution of Maccone

$$\sigma^2 \equiv \sum_J \left(1 - \frac{\mathbf{A_j}\mathbf{B_j}\left(\ln \mathbf{B_j} - \ln \mathbf{A_j}\right)^2}{\left(\mathbf{B_j} - \mathbf{A_j}\right)^2}\right)$$

it is the so-called *logarithmic variance* of the lognormal distribution of Maccone

$$\Phi(\mathbf{X_0}) \equiv \frac{1}{\mathbf{X_0}} \bullet \frac{1}{\sqrt{2\pi}\sigma} e^{-\frac{(\ln(\mathbf{X_0}) - \mu)^2}{2\sigma^2}}$$

It is the Maccone lognormal distribution function of the overall probability $\mathbf{X_0}$

$$\Delta \mathbf{T}_0 = \sum_j \Delta \mathbf{T}_{0j}$$ it is the total time of risk conditions sum of the individual $\Delta \mathbf{T}_{0j}$

At this point, from the three input values for each phase $\mathbf{A}_j$, $\mathbf{B}_j$, and $\Delta \mathbf{T}_{0j}$, applying the lognormal formula, we would get the average $\mathbf{<X_0>}$ and the variance $\boldsymbol{\sigma}(\mathbf{X}_0)$ of the random variable $\mathbf{X}_0$. However, these are the average and the deviation of the probability of overcoming ALL the bottlenecks in the total time $\Delta \mathbf{T}_0$, while now we want, given an appropriate confidence value of the survival probability **p** (for example the **50%**), the average and the deviation of another random variable which is the survival time $\Delta \mathbf{T}_p$ of the civilization.

We then proceed in the following way to obtain the distribution with respect to the random variable $\Delta \mathbf{T}_p$ function of $\mathbf{X}_0$; let's assume:

$$\frac{\Delta T_p}{\Delta T_0} \equiv \delta$$

$\boldsymbol{\delta}$ is the fraction of time relative to $\Delta \mathbf{T}_0$ for which we impose our survival probability **p**. Obviously, the survival probability to the *challenge* (*contravariant* with respect to time) in time $\Delta \mathbf{T}_p$ will be:

$$< \mathbf{X}_0 >^{\delta} = p$$

Extracting the natural logarithm and isolating $\Delta \mathbf{T}_p$ we get:

$$\Delta T_p = \left(\sum_j \Delta T_{0j} \right) \bullet \frac{\ln p}{\ln \langle X_0 \rangle} = \Delta T_0 \bullet \frac{\ln p}{\ln \langle X_0 \rangle}$$

Now, for convenience, let's set:

$$\begin{cases} -\Delta T_0 \ln \mathbf{p} = C > 0 \\ \Delta T_p = y > 0 \end{cases}$$

The previous equation can be succinctly written as:

$$\mathbf{y} = -\mathrm{C} / \ln(\mathrm{X}_0)$$

Note that we have moved from the average value $\mathbf{<X_0>}$ to the complete random variable $\mathbf{X}_0$. By reversing the last expression, we get:

$$\mathbf{X}_0 = e^{-\frac{C}{y}}$$

which is a monotonically increasing function. Then, according to the theorem of distribution of a function of a random variable, the new distribution for the variable $\mathbf{y} = \mathbf{\Delta T_p}$ is obtained from the lognormal distribution $\mathbf{\Phi(X_0)}$ by substituting the just found value of $\mathbf{X_0}$ as a function of $\mathbf{y}$ and multiplying everything by the derivative of $\mathbf{X_0}$ always as a function of $\mathbf{y}$. That is:

$$F_p(y) = \Phi\left(e^{-\frac{C}{y}}\right) \bullet \left(e^{-\frac{C}{y}} \bullet \frac{C}{y^2}\right)$$

Substituting the known values of $\mathbf{y}$ and $\mathbf{C}$, we have:

$$F_p(\Delta T_p) = \Phi\left(e^{\frac{\Delta T_0 \ln p}{\Delta T_p}}\right) \bullet e^{\frac{\Delta T_0 \ln p}{\Delta T_p}} \bullet \left(\frac{-\Delta T_0 \ln p}{\Delta T_p{}^2}\right)$$

And finally, in a more compact form:

$$F_p(\Delta T_p) = \Phi\left(p^{\frac{\Delta T_0}{\Delta T_p}}\right) \bullet p^{\frac{\Delta T_0}{\Delta T_p}} \bullet \left(\frac{\Delta T_0 \ln \frac{1}{p}}{\Delta T_p{}^2}\right)$$

This is our distribution function of the duration of galactic civilizations. Once the input values of all the phases/challenges $\mathbf{A_j}$, $\mathbf{B_j}$,, and $\mathbf{\Delta T_{0j}}$, are fixed, we will calculate $\mathbf{<\Delta T_p>}$ and $\mathbf{\sigma(\Delta T_p)}$, from the very definitions of mean and variance, using numerical methods. The curves shown in Fig. 20.1 demonstrate, for example, in box **A**, the classic Maccone lognormal function with respect to the fixed time value of $\mathbf{\Delta T_0 = 450,000}$ years determined by the sum of all critical phases; box **B** instead represents the distribution functions of the duration $\mathbf{\Delta T_p}$ of a galactic civilization, calculated from the Maccone lognormal function, and imposing survival probabilities of the civilization progressively from **40%** to **60%**: it is evident the decrease in time $\mathbf{\Delta T_p}$ as the chosen survival probability increases.

We verify this result by finding the location of the maximum points of the curves shown in Fig. 20.1b. We remember that the abscissas of these points are not exactly the average values of the curves because they are not perfectly symmetrical, but we can still make this approximation.

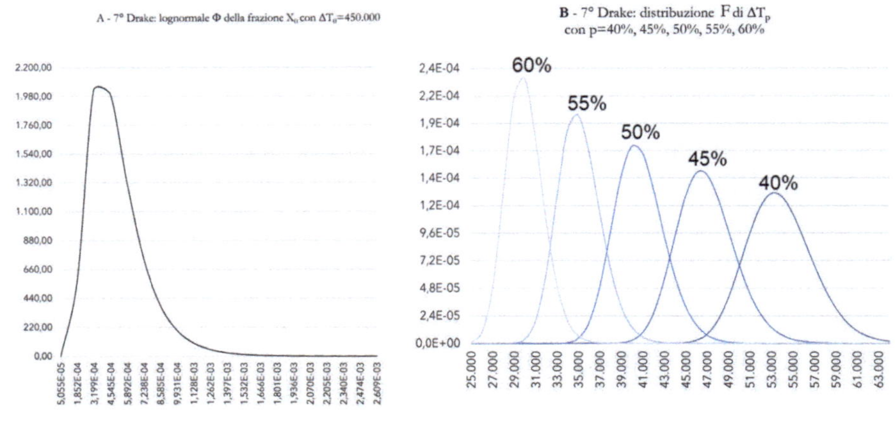

Fig. 20.1 Seventh Drake—(**a**): Classic Maccone lognormal function; (**b**): distribution functions of the duration **ΔTp (IJA 14/06/2023 Mieli, Valli, Maccone)**

First, we calculate the derivative with respect to $\mathbf{\Delta T_p}$ of our distribution function $\mathbf{F(\Delta T_p)}$. The calculation is a bit laborious, but after a few steps (APPENDIX **C**), we will obtain the derived function shown below:

$$\frac{d}{d\Delta T_p}F\left(\Delta T_p\right)=-\Phi\left(\boldsymbol{p}^{\frac{\Delta T_0}{\Delta \boldsymbol{T}_p}}\right)\bullet \boldsymbol{p}^{\frac{\Delta T_0}{\Delta \boldsymbol{T}_p}}\bullet\left(\frac{\Delta T_0 \ln\frac{1}{\boldsymbol{p}}}{\boldsymbol{\sigma}^2\Delta \boldsymbol{T}_p{}^4}\right)$$

$$\bullet\left[\left(\ln\left(\frac{1}{p}\right)\frac{\Delta T_0}{\Delta \boldsymbol{T}_p}+\mu\right)\ln\left(\frac{1}{p}\right)\Delta T_0-2\Delta \boldsymbol{T}_p\boldsymbol{\sigma}^2\right]$$

Setting the condition of the stationary point of the derivative (we do not report, for brevity, the condition on the second derivative **<0**):

$$\frac{d}{d\Delta T_p}F\left(\Delta T_p\right)=0$$

we obtain the second degree equation with $\mathbf{y} = \mathbf{\Delta T_p}$:

$$\left(-2\sigma^2\right)y^2+\left(\mu y_0\, ln\left(\frac{1}{p}\right)\right)y+\left(y_0\, ln\left(\frac{1}{p}\right)\right)^2=0$$

Which has a unique solution for positive values of $\mathbf{\Delta T_p}$.

The parametric coordinates, with respect to the parameter **p**, of the maximum points are then:

$$\begin{cases} \Delta T_{pMAX} = \boldsymbol{C} \bullet \ln\left(\dfrac{1}{p}\right)\Delta T_0 \\ F\left(\Delta T_{pMAX}\right) = \boldsymbol{D} \bullet \dfrac{1}{\ln\left(\dfrac{1}{p}\right)\Delta T_0} \end{cases}$$

and, consequently, from the first of the two:

$$\begin{cases} p = e^{-\frac{\Delta T_{pMAX}}{\tau}} \\ \tau = \boldsymbol{C} \bullet \Delta T_0 \end{cases}$$

having set the positive constants:

$$\begin{cases} \boldsymbol{C} = \left(\dfrac{\sqrt{\mu^2 + 8\sigma^2} + \mu}{4\sigma^2}\right) > 0 \\ \boldsymbol{D} = \dfrac{\exp\left[-\dfrac{1}{2\sigma^2}\left(\dfrac{4\sigma^2}{\sqrt{\mu^2 + 8\sigma^2} + \mu} + \mu\right)^2\right]}{\sqrt{2\pi}\sigma\left(\dfrac{\sqrt{\mu^2 + 8\sigma^2} + \mu}{4\sigma^2}\right)^2} > 0 \end{cases}$$

The product $\boldsymbol{C} \cdot \boldsymbol{D} = \boldsymbol{cost} > \boldsymbol{0}$, therefore:

$$\Delta T_{pMAX} \bullet F\left(\Delta T_{pMAX}\right) = \boldsymbol{C} \bullet \boldsymbol{D} == \dfrac{\exp\left[-\dfrac{1}{2\sigma^2}\left(\dfrac{4\sigma^2}{\sqrt{\mu^2 + 8\sigma^2} + \mu} + \mu\right)^2\right]}{\sqrt{2\pi}\sigma\left(\dfrac{\sqrt{\mu^2 + 8\sigma^2} + \mu}{4\sigma^2}\right)} = \boldsymbol{cost}$$

Therefore, the location of the maximum points of our time distributions will be a branch of an equilateral hyperbola as we expected (Fig. 20.2a).

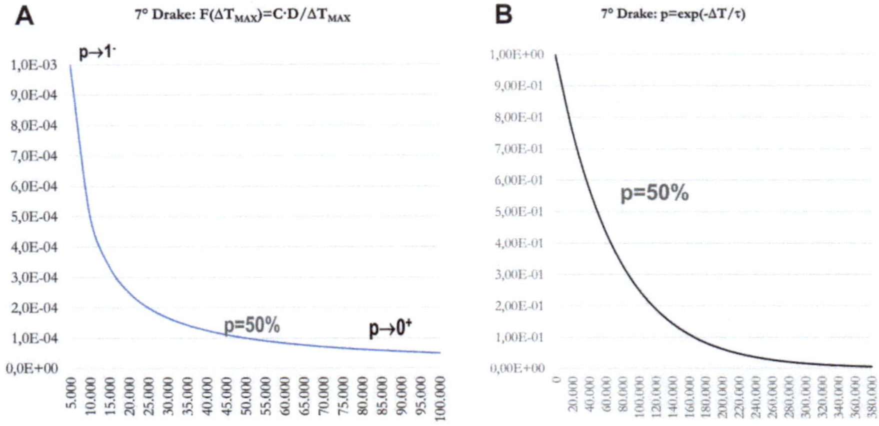

Fig. 20.2 (**a**) Seventh Drake: The equilateral descent of the maximum of the duration distribution functions of civilization **(IJA 14/06/2023 Mieli, Valli, Maccone)**. (**b**) Seventh Drake: The exponential descent of the probability of survival over time **ΔT** with a τ = **72, 300** years approximately **(IJA 14/06/2023 Mieli, Valli, Maccone)**

This means that, analyzing the two trivial limit cases, the certainty of survival ($\mathbf{p} \rightarrow \mathbf{1}^-$ or the probability that tends to **1** from the left) makes our time distribution function diverge into a Dirac delta at the origin of time, which in layman's terms means that only the present is certain, while for infinitesimal probabilities ($\mathbf{p} \rightarrow \mathbf{0}^+$ or the probability that tends to 0 from the right) both the duration of civilization and its deviation diverge, which means that the civilization could last an arbitrary time. Only the intermediate cases, for example ($\mathbf{p} = \mathbf{0}, \mathbf{5}$), give us useful information for our distribution of lifetimes. The Fig. 20.2b is, trivially, the exponential descent of the probability of survival over time $\mathbf{\Delta T}$ with, for example, $\boldsymbol{\tau}$ = 72, **300**.

20.3 The Seven Challenges of Galactic Civilizations (and the Plan B)

Now we need to solve two difficulties:

(A) imagine future scenarios even remote in space and time
(B) decouple such scenarios from our planetary and cultural context

In other words: what challenges could ANY evolving galactic civilization likely face to endure over time? We hypothesized the following seven risk factors and a Plan B for each:

1. self-destruction due to evolutionary insufficiency
2. self-destruction due to a technological error
3. technological insufficiency to cope with planetary changes
4. spontaneous natural involution of civilization
5. artificial genetic involution of civilization
6. robotic transition ended on a dead track
7. reaching **point Ω** and subsequent isolation of evolved intelligence

PLAN B—interstellar travel in case of failure

The first three risks are familiar, as they are typical of a technological civilization in its dawn like ours and are hotly debated today. The remaining four are scenarios that could occur once the first three are overcome; they may seem less fearsome because they don't (or don't seem to) be imminent, but reasoning in terms of tens of thousands of years, they constitute threats as devastating as the first ones.

Interstellar travel is the potential escape route from a hypothetical failure with the different challenges. Since the challenges present themselves in more or less early periods of civilization, interstellar travel is an effective escape or not depending on the challenge it is applied to.

20.4 The First Challenge: Self-Destruction Due to Evolutionary Insufficiency

It is inevitable that an intelligent species equipped with the instinctual and cultural baggage that supported it in its evolutionary journey will, at a certain point, face a sudden technological acceleration. The reason is obvious: science, but often nature as well, does not progress gradually but proceeds in sudden and unexpected leaps, followed by periods of relative stasis. These accelerations are a risk (Fig. 20.3).

In our previous paragraph on Drake's fifth parameter, we followed Kardashev's footsteps to catalog the level of a galactic civilization on four levels:

$\mathbf{W}_1 = 10^{16}$ Watt	is all the solar power that a rocky planet orbiting in its habitability zone receives
$\mathbf{W}_2 = 10^{11}\cdot\mathbf{W}_1$	is all the power radiated by the star
$\mathbf{W}_3 = 10^{11}\cdot\mathbf{W}_2$	is all the power radiated by the galaxy
$\mathbf{W}_4 = 10^{11}\cdot\mathbf{W}_3$	is all the power radiated by the observable universe

Fig. 20.3 Outcome of the evolutionary insufficiency to manage energy Powered by OpenAI

According to this scale, the level of a civilization, which develops a total power of $\mathbf{W}_{ETC}$, is expressed by the index $\mathbf{k} = \mathbf{1}, \mathbf{2}, \mathbf{3}$ and $\mathbf{4}$ in the subscript of the powers $\mathbf{W}_K$, and it is calculated:

$$\boldsymbol{K} = \frac{log_{10}\left(W_{ETC}\right) - 5}{11}$$

For a common animal species (including humans until two centuries ago), **K** did not exceed **0.6**, but for us today, it already stands at **0.7** and is continuously growing. This means our access to energy is increasing, and simultaneously, the level of responsibility we must have in managing this level of energy is growing. In a nutshell, a generic galactic civilization must try not to be consumed by the fire it has discovered, especially if the energy levels involved are planetary, as is atomic energy in our current case.

However, a galactic civilization does not automatically have such capabilities within its heritage of instincts, simply because technological changes belong only to the last phase of its development: culture. We can only hope that culture can channel behaviors, derived from the natural base that shaped the species, in the right direction over time. It is a difficult task, perhaps desperate, but the stakes are high: survival in the immediate future. This is what we must do in controlling nuclear weapons, and not only that.

We believe it is plausible that this passage is universal for galactic civilizations and that it is very difficult to overcome (just around **10%** succeed). The challenge arises from the moment a civilization is potentially able to express a technology and ends when the society's culture has defused any lingering instincts from the natural past that can lead to self-destruction (about 100,000 years). The probability of overcoming the challenge is between 0.05 and 0.15 over a challenge time limit of 100,000 years.

CHALLENGE 1

A_1	B_1	ΔT_{01}
0.05	0.15	100,000

20.5 The Second Challenge: Self-Destruction Due to a Technological Error

This second case resembles the first, but lacks *malice*—even though it still depends on poor management of acquired technology, self-destruction is due to a misjudgment where neither the nature nor the culture of a civilization bears responsibility. For example, a nuclear war could occur due to a mere misunderstanding, or a disease could unintentionally spread from a research laboratory (Fig. 20.4).

It is a challenge that can overlap temporally, even partially, with the first but tends to arrive later when the acquired technological power is very high, the problems seem solved, and the civilization lowers its level of control over the potential consequences of its actions.

This passage is also inevitable for galactic civilizations and quite difficult to overcome (around **20%** succeed). The challenge presents itself from the moment a civilization possesses mature technology and ends when the technology itself provides sufficient tools to put the civilization in safety, such as interstellar travel capabilities. The probability of overcoming this challenge is between 0.1 and 0.3 over a 50,000-year time limit.

Fig. 20.4 Outcome of unforeseen technological errors Powered by OpenAI

CHALLENGE 2

A_2	B_2	ΔT_{02}
0.1	0.3	50,000

20.6 The Third Challenge: Technological Insufficiency to Face the Planetary Changes That Have Occurred

This phase strictly depends on Drake's third parameter, which gives the probability for Earth-like planets to remain stably in their habitable zone. Since the third parameter also has a statistical distribution, we must consider planets

Fig. 20.5 Outcome of unforeseen planetary changes Powered by OpenAI

subjected to traumas, even if not definitive, but still significant disruptions to their stability during a civilization's existence. Known examples are recurring meteor bombardments, possible significant shifts of the planet's rotational axis, and severe climate changes due to geological activity (Fig. 20.5).

In this case, the only countermeasure a civilization has to face such events is an adequate technological level, let's say the achievement of a Kardashev parameter between **1** and **2**, which is no small feat.

This step is less common than the previous ones for galactic civilizations (since it depends on the third parameter) and is easier to overcome (around **50%** succeed). Also in this case, the challenge presents itself from the moment a civilization possesses mature technology and ends when the technology itself provides sufficient tools to put the civilization in safety. The probability of overcoming this challenge is between 0.4 and 0.6 over a 20,000-year time limit.

CHALLENGE 3

A_3	B_3	ΔT_{03}
0.4	0.6	20,000

20.7 The Fourth Challenge: Spontaneous Involution

With this parameter, we move into the true future—into scenarios less familiar to science and more explored by science fiction. The fact that a civilization must necessarily progress technologically is certainly our conviction, dictated by the continuous presence of environmental pressures, often caused by ourselves, which favors a technologically advanced civilization over a less advanced one, WHATEVER THE PRICE TO PAY. Even in our small terrestrial civilization, we have learned that this is not always true, or rather, it is not possible to talk about technological progress without considering the price to pay (Fig. 20.6).

Obviously, we do not think galactic civilizations are massively exposed to a *flower children* syndrome, but in cases where external and internal dangers

Fig. 20.6 Involved alien community Powered by OpenAI

seem less imminent, the choice not to progress, and therefore not to risk, is possible. In those cases, the civilization could deliberately choose never to become an evolved ETC (extra-terrestrial civilization) with a high Kardashev parameter, but to live comfortably with its own **K** between **0.6** and **0.7** without much trauma and little chance of being identified.

There is also the case that this involution phenomenon is determined by the progressive robotization of activities at the expense of the initiatives of the civilization that originally created those automatisms. In short, if machines do everything, what is the purpose of changing, evolving, exploring and, above all, risking? A civilization could inadvertently stagnate and remain stagnant [42].

We also consider this an uncommon step for galactic civilizations that is easy to overcome (around **80%** succeed). The challenge arises when a civilization possesses mature enough technology to make it feel safe and ends when it has chosen to expand beyond its own planet: the probability of overcoming the challenge is between 0.7 and 0.9 over a challenge time limit of 30,000 years

CHALLENGE 4

A_4	B_4	ΔT_{04}
0.7	0.9	30,000

20.8 The Fifth Challenge: The Artificial Genetic Transition Ended on a Dead Track

A few years ago, we could have treated this parameter as one of the remote science fiction scenarios. However, we must note that in 2014 the article "The new frontier of genome engineering with CRISPR-Cas9" by Jennifer A. Doudna and Emmanuelle Charpentier [20] was published, effectively opening the way to precision genetic manipulation (Fig. 20.7).

This means the evaluation of this scenario overlaps with the first scenario, which instead deals with the risk of self-destruction due to limits inherent to the species (less elegantly said, *stupidity*). We therefore omit the part related to the first parameter, as it has already been dealt with, focusing only on the long-term consequences of a potential negative drift resulting from genetic manipulation. The negative consequences could be as follows: unlike natural selection which takes place over geological periods and gives the ecosystem time to always be in balance, artificial genetic manipulation is decided rapidly, without the certainty that the ecosystem will rebalance. A superficial use of these technologies (we are not talking about us earthlings who will certainly use it superficially) could lead to unwanted consequences in the medium and long term, such as a progressive reduction in the number of individuals due to accidental infertility or similar phenomena.

This step is probably common for galactic civilizations and quite easy to overcome (around **70%** succeed). The challenge arises from the moment a civilization possesses sufficiently sophisticated technology (for us, it is today!) and ends when such technology is at a high enough level that it can correct itself if necessary. The probability of overcoming this challenge is between 0.5 and 0.8 over a 50,000-year time limit.

CHALLENGE 5

A_5	B_5	ΔT_{05}
0.5	0.8	50,000

20.9 The Sixth Challenge: Transition of Finite Artificial Intelligence onto a Dead-End Track

We cannot address this parameter without defining what we mean by Artificial Intelligence. For this purpose, we will follow the steps of Roger Penrose with his monumental works "The Emperor's New Mind" [86] from 1990, and "Shadows of the Mind" [88] from 1994, which are cornerstones on this topic (Fig. 20.8).

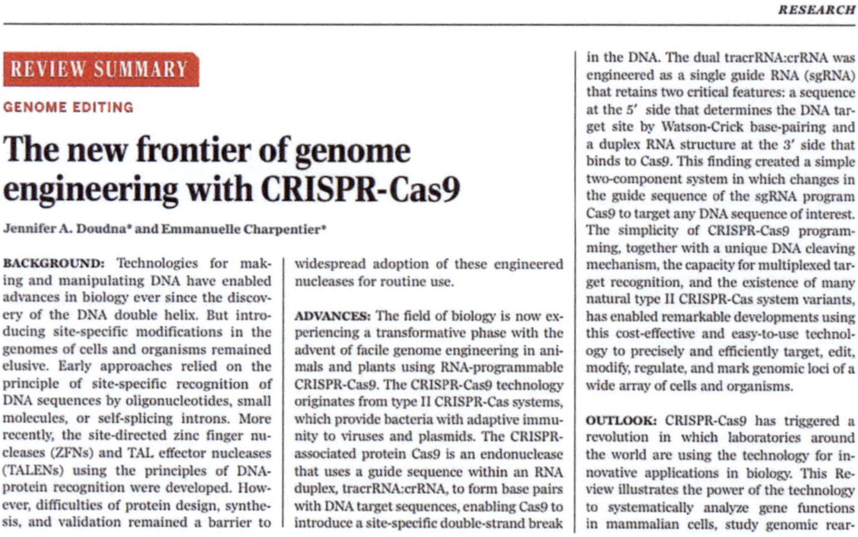
RESEARCH

REVIEW SUMMARY

GENOME EDITING

The new frontier of genome engineering with CRISPR-Cas9

Jennifer A. Doudna* and Emmanuelle Charpentier*

BACKGROUND: Technologies for making and manipulating DNA have enabled advances in biology ever since the discovery of the DNA double helix. But introducing site-specific modifications in the genomes of cells and organisms remained elusive. Early approaches relied on the principle of site-specific recognition of DNA sequences by oligonucleotides, small molecules, or self-splicing introns. More recently, the site-directed zinc finger nucleases (ZFNs) and TAL effector nucleases (TALENs) using the principles of DNA-protein recognition were developed. However, difficulties of protein design, synthesis, and validation remained a barrier to widespread adoption of these engineered nucleases for routine use.

ADVANCES: The field of biology is now experiencing a transformative phase with the advent of facile genome engineering in animals and plants using RNA-programmable CRISPR-Cas9. The CRISPR-Cas9 technology originates from type II CRISPR-Cas systems, which provide bacteria with adaptive immunity to viruses and plasmids. The CRISPR-associated protein Cas9 is an endonuclease that uses a guide sequence within an RNA duplex, tracrRNA:crRNA, to form base pairs with DNA target sequences, enabling Cas9 to introduce a site-specific double-strand break in the DNA. The dual tracrRNA:crRNA was engineered as a single guide RNA (sgRNA) that retains two critical features: a sequence at the 5′ side that determines the DNA target site by Watson-Crick base-pairing and a duplex RNA structure at the 3′ side that binds to Cas9. This finding created a simple two-component system in which changes in the guide sequence of the sgRNA program Cas9 to target any DNA sequence of interest. The simplicity of CRISPR-Cas9 programming, together with a unique DNA cleaving mechanism, the capacity for multiplexed target recognition, and the existence of many natural type II CRISPR-Cas system variants, has enabled remarkable developments using this cost-effective and easy-to-use technology to precisely and efficiently target, edit, modify, regulate, and mark genomic loci of a wide array of cells and organisms.

OUTLOOK: CRISPR-Cas9 has triggered a revolution in which laboratories around the world are using the technology for innovative applications in biology. This Review illustrates the power of the technology to systematically analyze gene functions in mammalian cells, study genomic rear-

Fig. 20.7 The paper of nobel laureates Jennifer A. Doudna and Emmanuelle Charpentier, discovers of CRISPR-Cas9

Artificial intelligence is not mere automation mentioned in the fourth challenge (robotization), but the actual birth of synthetic consciousness. Penrose defines simple automation, which includes all machines built to date, as **W**eak **A**rtificial **I**ntelligence (**WAI**), while the possibility of creating conscious machines is defined as **S**trong **A**rtificial **I**ntelligence (**SAI**). The power of Penrose's reasoning lies in demonstrating that the second CANNOT be a simple extension of the first—that machines, no matter how complex, BASED ONLY ON ALGORITHMIC FUNCTIONING cannot become conscious.

The demonstration is complex, but worth a brief mention: through Gödel's incompleteness theorem of 1931 [36], Penrose points out that the possibility of formulating an unprovable but true theorem, like the one constructed by Gödel, clearly demonstrates the impossibility for a machine to do the same. The theorem states that it is possible to construct an infinite number of TRUE but UNPROVABLE theorems through a starting formal system and finite number of axioms; this system would function like an algorithm. How then does the human mind, if it functioned like an algorithm however complex, manage to formulate a theorem like Gödel's proof? The answer is that the human mind is conscious, and consciousness does NOT rely on an algorithm, but on something further. For Penrose, this is an aspect of still-controversial quantum mechanics, namely the collapse of the quantum wave function. This phenomenon would not occur according to the subjective Copenhagen interpretation mechanism (i.e., the analyzed system *decides* the measurement result at the moment of interaction with the observer), but according to a spontaneous, objective mechanism detached from the observer and linked to the energy level involved. The process would occur within the microtubules of neuronal

Fig. 20.8 *The emperor's new mind*, the main work of the Nobel laureate Roger Penrose, creator the absolute separation between Strong Artificial Intelligence (SAI) and Weak Artificial Intelligence (WAI)

cytoskeletons. As of October 2022, further confirmation of this line of thought was provided by Christian Matthias Kerskens and David López Pérez [53].

Without delving further into this topic, what interests us is noting that we are far from building conscious machines; we can at most build excellent and fast automatisms, nothing more. To build true conscious machines, we would have to equip them with a mechanism as effective as the one present in animal brains for the *objective collapse of the quantum wave function.*

All of this may seem abstract and naive, but Penrose formulated this hypothesis in 1990 when it was thought *Moore's law* (Fig. 20.9) on the exponential increase of computing power would have inevitably and soon led us to artificial intelligence. After more than 30 years we only have very fast machines that cannot make real evaluations without adequate instruction. So what are we discussing when we talk about the transition to artificial intelligence? Certainly not the control unit of the International Space Station or voice response systems, as *stupid* as a thermostat. We mean precisely that technological change

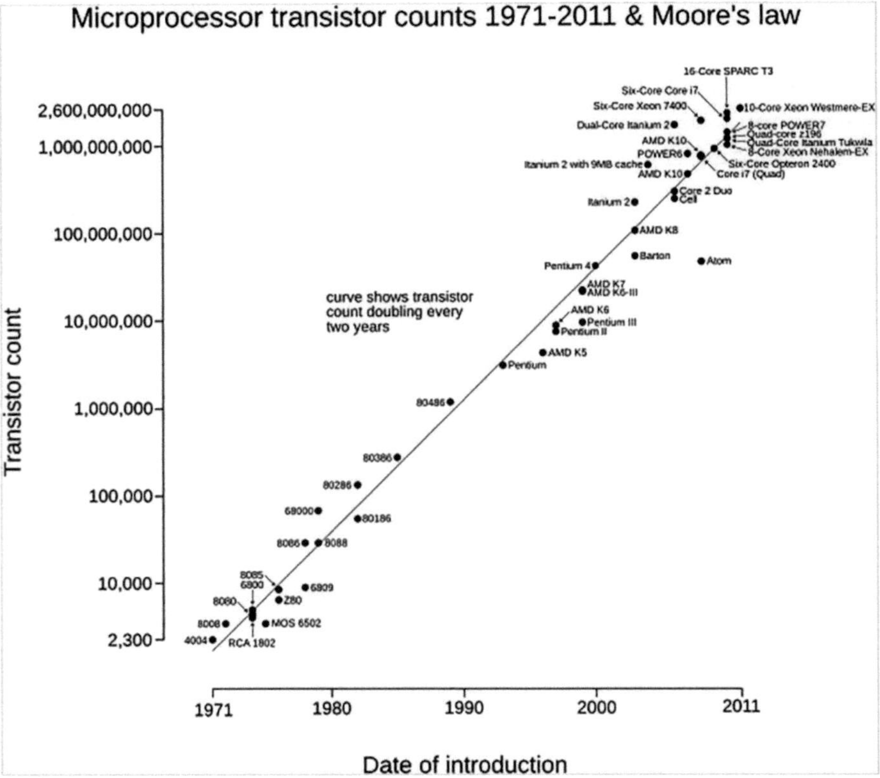

Fig. 20.9 Graph of Moore's law on the exponential increase of computing power over time (Wgsimon)

giving machines access to the true mechanism of animal consciousness. If this leap were to occur, then yes, a galactic civilization would have to face how to manage these new synthetic consciousnesses, in the same way as in the fifth challenge it had to face managing the artificial genetic transition.

We have learned that, while the genetic transition is already happening for us after the CRISPR-Cas9 discovery, the transition to SAI is still far away. But if it were to happen, what challenge would a galactic civilization face?

Science fiction is full of stories on the subject, summarized in one sentence: a civilization would find itself having to manage the emergence of a new synthetic species that potentially would have the means to dominate it. To avoid this danger, *the newborn synthetic consciousness (SAI) should be rigidly separated from the extraordinary technological tools that instead come from automation (WAI)*. The challenge is this: avoiding putting a nuclear bomb in the hands of a conscious machine. In case of failure, the galactic civilization could be destroyed by the emerging synthetic civilization, which then might not have the tools or motivations to survive itself (after all, what drives a synthetic consciousness?).

This transition has the same probabilistic and temporal characteristics as the previous one (it doesn't matter if the first is happening to us and the other is not yet): the probability of overcoming the challenge is between 0.5 and 0.8 over a 50,000-year time limit.

CHALLENGE 6

A_6	B_6	ΔT_{06}
0.5	0.8	50,000

20.10 The Seventh Challenge: Reaching the point Ω

If the sixth challenge of transitioning to Strong Artificial Intelligence (SAI) were overcome, not all problems would necessarily be solved. In 1993, the writer and mathematician Vernor Vinge hypothesized the occurrence of a phenomenon called the "singularity" or **point Ω** [113]. This refers to a real explosion of intelligence due to transferring its development principle from organic to synthetic bases, resulting in an acceleration of the improvement trend—in short, a civilization designs and builds intelligent machines that in turn design and build smarter machines, and so on (Fig. 20.10).

It's easy to realize that, if the problem of realizing SAI and containing it safely were solved, the destiny of a civilization could be the **point Ω**. The

Fig. 20.10 The **point Ω** is the unstoppable progress of artificial consciousness, once it has formed and acquired autonomy to improve itself Powered by OpenAI

characteristic of this point or singularity is precisely to go beyond our logical-scientific and moral understanding. Like a black hole, of **point Ω** we only know that intelligence expands at that point, and the decisions made by such a civilization are inaccessible to us. One possible decision is to permanently exit the radar of *primitive* civilizations (like ants to us) as **NOT Ω** and make themselves completely invisible.

We cannot exclude the possibility of a galactic civilization reaching such an exotic outcome, but as already said, we still know too little about the mechanisms of SAI to conclude **point Ω** is inevitable. For this reason, overcoming the risk of point **Ω** is attributed a high probability of success between 0.5 and 0.9 over a 100,000 year time limit.

CHALLENGE 7

A_7	B_7	ΔT_{07}
0.5	0.9	100,000

Table 20.1 Seventh Drake: The π_j probabilities of success of the **B** plans related to each of the seven challenges of the seventh parameter

j	Challenge	π_j
1	Self-destruction due to evolutionary insufficiency	1%
2	Unintentional technological error	3%
3	Technological insufficiency to face planetary changes	5%
4	Spontaneous involution	10%
5	Artificial genetic transition ended on a dead track	20%
6	Transition of artificial intelligence ended on a dead track	20%
7	Reaching point Ω	50%

20.11 Plan B: Escape to Other Planets and Interstellar Travel

It is reasonable to think that, in case of announced failure in one or more of the previous challenges, galactic civilizations would try to move to other solar systems deemed suitable. It should be emphasized that these are not planned trips, but dictated by emergencies; therefore, their probability of success is low, especially if the galactic civilization is not mature enough [68]. To take into account the maturity of the civilization at the time it might attempt interstellar travel, we have assigned variable and gradually increasing probabilities $\boldsymbol{\pi}_j$ as the hypothetical technological baggage present at the time of the challenge in question increases.

The minimum and maximum probabilities $\mathbf{A}_j$ and $\mathbf{B}_j$, previously defined, are then corrected through the following formula:

$$\mathbf{A}_j' = \mathbf{A}_j + \left(\mathbf{1} - \mathbf{A}_j\right) \cdot \pi_j$$

$$\mathbf{B}_j' = \mathbf{B}_j + \left(\mathbf{1} - \mathbf{B}_j\right) \cdot \pi_j$$

In this way the values $\mathbf{A}_j'$ and $\mathbf{B}_j'$, which take into account the **B** plan of interstellar displacement, are slightly increased compared to $\mathbf{A}_j$ and $\mathbf{B}_j$. The assigned probabilities $\boldsymbol{\pi}_j$ are the following (Table 20.1):

20.12 The Calculation of the Seventh Drake Parameter

At this point we have all the elements to calculate the duration of a galactic civilization. The Table 20.2 and the Fig. 20.11a, b show the calculation carried out starting from the four input values for each challenge. The

Table 20.2 Seventh Drake: The **28** input values of the seventh parameter

	Minimum probability over time ΔT_{0j}	Maximum probability over time ΔT_{0j}	Maximum time challenge j (years)	Chance of success PLAN B
Challenge	A_j	B_j	ΔT_{0j}	π_j
1	0.05	0.15	100,000	1%
2	0.10	0.30	50,000	3%
3	0.40	0.60	20,000	5%
4	0.70	0.90	30,000	10%
5	0.50	0.80	50,000	20%
6	0.50	0.80	50,000	20%
7	0.50	0.90	100,000	50%

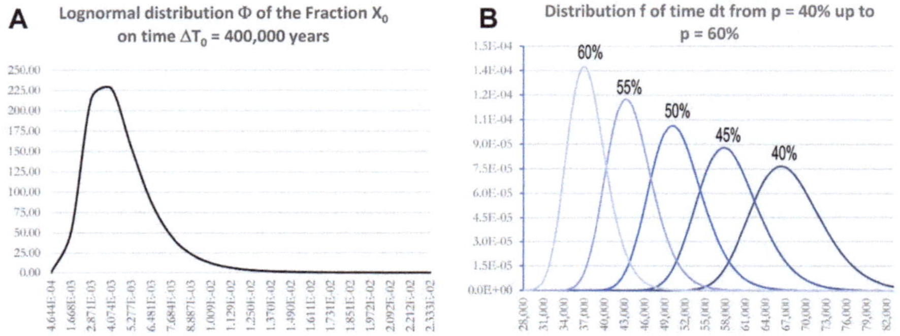

Fig. 20.11 **(a)** Seventh Drake: Maccone's lognormal **Φ** of the probability $\mathbf{X_0}$ to overcome the entire process $\mathbf{\Delta T_0}$ **(IJA 14/06/2023 Mieli, Valli, Maccone)**. **(b)** Seventh Drake: Distribution **F** of the duration **ΔT** for five values of the reference probability **p (IJA 14/06/2023 Mieli, Valli, Maccone)**

resulting parameter $\mathbf{f_L}$ is nothing more than the ratio between the found duration (about **50,000 years**) and the duration of the last stellar population (**7 Ga**):

$$\mathbf{f_L = 7.3 \cdot 10^{-6}}$$

DRAKE 7 (static civilization)

$\mathbf{f_{L\,min}}$	$\mathbf{f_{L\,max}}$
$3 \cdot 10^{-6}$	$8.3 \cdot 10^{-6}$

To be fair, we must admit that up until now we have hypothesized that the probability $\mathbf{p_0}$ calculated with the lognormal distribution on the total time $\mathbf{\Delta T_0}$ was approximately independent of time. Therefore, as we have seen, its extrapolation to time $\mathbf{\Delta T}$ follows a simple exponential law of the form:

$$p = \exp\left(-\Delta T / \tau\right)$$

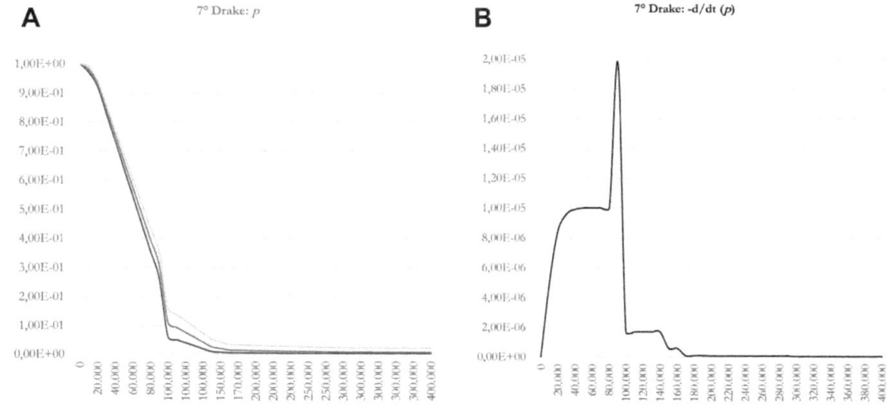

Fig. 20.12 (a) Seventh Drake: The actual decrease of the survival probability over time ΔT for the case of the seven challenges **(IJA 14/06/2023 Mieli, Valli, Maccone)**. (b) Seventh Drake: The actual distribution function of the survival probability over time ΔT for the case of the seven challenges **(IJA 14/06/2023 Mieli, Valli, Maccone)**

However, this is actually NOT TRUE, especially for values of $\mathbf{\Delta T}$ less than $\mathbf{\Delta T_0}$, because the probability is nothing more than the partial combination of the different probabilities of the individual seven challenges. This leads to a more complex curve for the **p** function and for the probability density distribution of **p** (the latter being merely the changed derivative of **p** sign), as shown in Fig. 20.12a, b. Nevertheless, calculating the seventh Drake parameter using this more laborious method would lead us to a result on the same order of magnitude as the one we found assuming a probability strictly independent of time.

21

Considerations on the Seventh Parameter

As can be seen from Fig. 20.11b, we have found an average duration of galactic civilizations of about **50,000** years with a deviation of **7000** years. Now **assuming** a reasonable duration of a planet's life in the habitable zone of about 7 **billion** years (**Gy**) (remember that the third Drake parameter assures us this), the value of the temporal fraction of a civilization's life is $\mathbf{7.3 \times 10^{-6}}$ with a deviation of $\mathbf{10^{-6}}$. *This means that our galactic civilization has a total time of about* ***50,000*** *years, starting from our current technological level, during which it can recognize and be recognized by other civilizations.*

Let's briefly return to the delta-T reasoning, which, as you will remember, we did not consider applicable: Gott had found a value for the duration of the **HUMAN SPECIES** between **70,000** and **600,000** years with a probability of **50%**. Now, however, we have found a value of the duration of the **GALACTIC CIVILIZATION** between **44,000** and **58,000** years with a **50%** probability; we therefore have a definitely more accurate data and, above all, one that is referred to what interests the Drake equation, namely galactic technological civilizations rather than intelligent galactic species. If we apply the delta-T reasoning to advanced human technological civilizations (for example, since 1945, the beginning of the atomic era), we should seriously worry because we would have obtained a value for the duration of human technological civilization between **25** and **240** years; that is, we will have the probability of destroying ourselves, with a **50%** probability, between the years **2050** and **2265**. This would be a concrete possibility if two conditions were met: (A) the current moment was not a particular moment within the atomic era, and (B) all past and future time intervals were equivalent because conditions (i.e., our behaviors) remained unchanged over time. **All this should make us reflect seriously**.

E. Mieli et al., *The Living Galaxy*, https://doi.org/10.1007/978-3-031-67324-5_21

22

The Hypothesis of Tipler and Brin of Dynamic Civilizations

At this point, a final consideration is mandatory: until now we have only considered the case where a civilization forms and develops on its own home planet and moves ONLY TO ESCAPE A POTENTIALLY CATASTROPHIC DANGER due to one of the seven challenges; we have thus ignored the scenario analyzed by Tipler (1980) and Brin (1983) of a civilization that, at a certain point in its development, freely decides to colonize the galaxy. How and when would such a scenario present itself?

First of all, it is conceivable that a civilization willing and able to colonize the galaxy is a type K2 civilization that has already overcome ALL seven challenges after **400,000** years; but in this case, we would be facing an epochal transition of the entire process: if this were to happen, the civilization in question COULD NO LONGER BECOME EXTINCT. Therefore, the value that we should consider in our formula is not the one derived from the temporal distribution with a **50%** probability of survival that leads us to an average duration of **50,000** years equal to a percentage fraction of **7.3×10^{-6}**, but directly from the value provided by Maccone's formula to overcome all seven challenges in **400,000** years, which is much greater, namely **5×10^{-3}** (Fig. 20.11a).

DRAKE 7 (dynamic civilization)

$f_{L\,min}$	$f_{L\,max}$
1.03×10^{-3}	8.25×10^{-3}

E. Mieli et al., *The Living Galaxy*, https://doi.org/10.1007/978-3-031-67324-5_22

All of this leads to a conclusion that we should have expected, namely: if we were to identify an extraterrestrial civilization, it would be much more likely to be an evolved K2 civilization moving in the galaxy rather than a static K1 (or lower) civilization like ours still grappling with the challenges to overcome. In conclusion, a potential transition from K1 to K2, as hypothesized by Kardashev in 1964, is crucial and has a decisive impact on the calculation of the entire Drake equation.

Part IV

The Complete Drake Equation

At this point we have all seven of Drake's parameters to estimate galactic civilizations. To be precise, the fifth parameter has been explained in its three components relating to eukaryotes, metazoans, and technological intelligent civilization (ETC).

As we have just seen in the previous section, technological intelligent civilizations can be of two types: static (point **7 A** of **Table IV.1**) or dynamic (point

Table IV.1 TOTAL Drake: Drake's parameters defined with their minimum and maximum values. The fifth parameter has been explained in its three components related to eukaryotes, metazoans and technological intelligent civilization

		Minimum value	Maximum value
		A_j	B_j
1	Number of galaxy stars suitable for life (of spectral class F, G and K)	1.00E+10	1.20E+10
2	Number of suitable planets in the habitable zone per star (of spectral class F, G and K)	1.60E-01	2.00E-01
3	Fraction of stable planets for 4.5 Gy [**a**]	3.65E-02	1.07E-01
4	Fraction of planets where life arises	3.22E-01	6.55E-01
5	Eukaryotes	2.89E-01	7.09E-01
	Metazoans	5.97E-01	9.44E-01
	Technological intelligent civilization	1.25E-02	5.66E-02
6	Fraction of planets where life decides to communicate	4.00E-01	6.00E-01
7A	Temporal fraction of the duration of a static K1 type civilization	6.29E-06	8.30E-06
7B	Fraction of planets where life reaches the state of dynamic K2 civilization	1.03E-03	8.25E-03

[a]We have rescaled the value of the third parameter from **7 Gy** to **4.5 Gy** which is the specific duration of planet earth

7 B of **Table IV.1**); the former are those that have NOT made the leap beyond the level **K** = **1.4** and therefore have almost certainly NOT had the technological possibility of colonizing other planets; the latter instead have made the leap beyond the level **K** = **1.4** and therefore have spread beyond their original planetary system. We have obtained these two distinct results from a reflection born from the conclusions drawn from the 7th parameter. It is immediate, from simple energy calculations related to the power cost of an interstellar spaceship, to understand why this limit value of **K** = **1.4** was chosen: to accelerate a significant spaceship, let's say of $\mathbf{10^9\ kg}$, to half the speed of light, or $\mathbf{1.5 \cdot 10^8\ m/s}$ in about three months (approximately $\mathbf{10^7\ s}$), a power of about $\mathbf{10^{18}\ W}$ is required. Assuming that this is reasonably **1%** of the total power expressible by the civilization in question, this leads us to a total power of $\mathbf{W_{tot}} = \mathbf{10^{20}\ W}$, or:

$$\mathbf{K} = \frac{\log_{10} \mathbf{10^{20}} - 5}{11} \cong \mathbf{1,4}$$

This implies that only civilizations that have exceeded this power level can begin interstellar travel and colonization. The result obtained for static and dynamic civilizations is as shown in Table IV.2 and Fig. IV.1a, b.

At this point, we leave, for a moment, the discussion on galactic civilizations to show the entire set of results that have been obtained from this method.

Table IV.2 TOTAL Drake: number of static and dynamic civilizations

Average ETC number	Deviation ETC number	
<N>	σ(N)	
3	2	Static
2.000	2.000	Dynamic

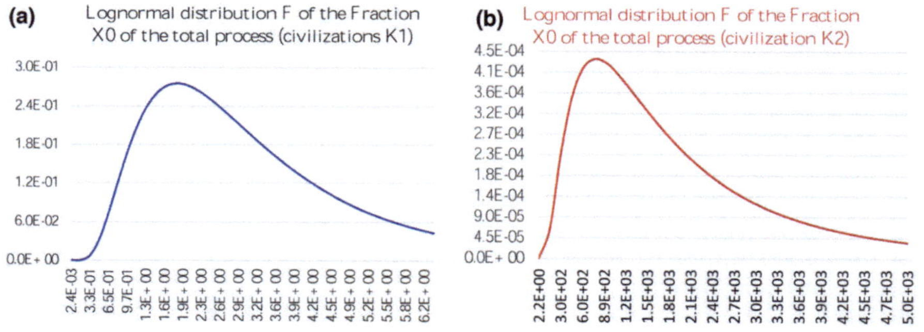

Fig. IV.1 **(a)** TOTAL Drake: lognormal of the distribution of static civilizations with **<N >** = **3.41** and **σ(N) = 2.43 (IJA 14/06/2023 Mieli, Valli, Maccone). (b)** TOTAL Drake: lognormal of the distribution of dynamic civilizations with **<N > = 2, 214** and **σ(N) = 2, 224 (IJA 14/06/2023 Mieli, Valli, Maccone)**

The third parameter, as we have seen, has given us the number of stable planets as a function of the variable time **ΔT**. Knowing now the duration of all the processes of life evolution, from the prokaryotic level to the K2 civilization, we can count:

A how many planets are suitable at each level of development
B how many of these have hosted life
C how many still host life

As we have seen, we must also consider the planet's development dead times (ADEANO), the first increase of oxygen to 1% of the current value (GOE), and the second increase to the current value (NOE). Combining these timings with the step-by-step calculation of the lognormal distribution, we get the time scale reported in Table IV.3, where we have adhered to the terrestrial model of life.

Now, taking into account the results obtained for the percentage of past and present stable planets for **ΔT** years (Fig. 6.1a, b) in the calculation of the third parameter, we report in Table IV.4 and Fig. IV.2 the population of

Table IV.3 TOTAL Drake: time scale of a galactic civilization (Earth model)

Life Development Phase	Duration	Sum
1 - ADEAN	0.80 Gy	0.80 Gy
2 - prokaryotes	**0.10** Gy	**0.90** Gy
3 - GOE	1.60 Gy	2.50 Gy
4 - eukaryotes	**0.50** Gy	**3.00** Gy
5 - NOE	0.50 Gy	3.50 Gy
6 - metazoans	**0.50** Gy	**4.00** Gy
7 - past static ETC	**0.50** Gy	**4.50** Gy
8 - present static ETC	**0.05** Gy	**4.55** Gy
9 – dynamic ETC	**0.40** Gy	**4.95** Gy

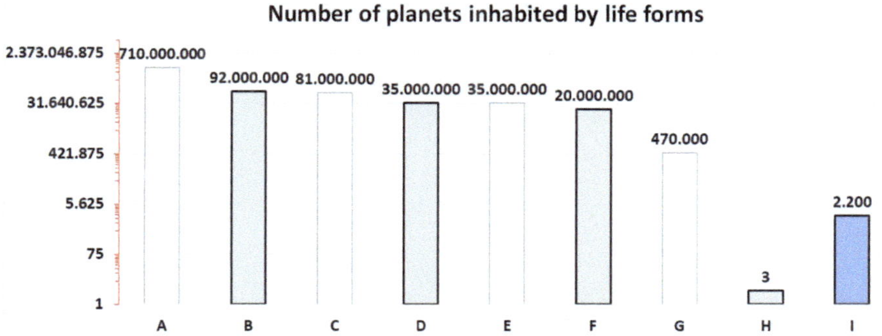

Fig. IV.2 TOTALE Drake: the data from Table 4 on the number of planets that have hosted or host life are reported in a logarithmic graph **(IJA 14/06/2023 Mieli, Valli, Maccone)**

Table IV.4 TOTAL Drake: the population of galactic life and the relative distances from us of both suitable planets and planets suitable and populated in the past and present: a average volume of the galactic disk of **1.53 × 10^{13} ly^3** has been hypothesized

	Planet age Gy)	Number of planets hosting or having hosted life	Distance (ly)	Total of suitable planets	Distance (ly)
A	0.90	710,000,000 planets where prokaryotes were born in the past	28	1,500,000,000	22
B		92,000,000 planets where prokaryotes are present today	55	190,000,000	43
C	3.00	81,000,000 planets where eukaryotes were born in the past	57	330,000,000	36
D		35,000,000 planets where eukaryotes are present today	76	140,000,000	48
E	4.00	35,000,000 planets where metazoans were born in the past	76	190,000,000	43
F		20,000,000 planets where metazoans are present today	91	110,000,000	52
G	4.50	470,000 planets where ETC K1 statistics were born in the past	319	140,000,000	48
H		3 planets where ETC K1 is present today static	16,515	92,000,000	48
I	4.95	2200 current planets with ETC K2 dynamics (eternal)	1909	92,000,000	48

galactic life and the relative distances from us. In **Appendix B** we have reported the complete calculation from which the data were derived.

In conclusion, we found that:

A The first major reduction occurs in life forms becoming ETC (**approximately 1/70 of the planets that host animal life forms**)

B A second dramatic reduction is due to the technological challenges of the 7th parameter (**approximately 1/250 of the ETCs that have developed**)

23

Still the Fermi Paradox (But Really, Where Is Everyone?)

After exploring ***fifty shades of Drake's equation***, spanning astronomy, geology, biology, paleontology, and futurology, we surprisingly found that the number of current planets inhabited by static type K1 (or lower) galactic civilizations, with an average duration of **50,000** years, is around **3**—that is, us and perhaps someone else about **17,000 light-years away**, which is very far.

If we only considered static civilizations or those that have not made the leap beyond **K = 1**, which would allow them to move among the stars, the Fermi paradox would, in fact, be solved [44]. To satisfy our desire for company in the galaxy, there would be a multitude of planets that host life in less evolved forms or that have hosted now-extinct civilizations – daily bread for astrobiologists or astroarchaeologists, **but nothing more than that …**

… but instead, there is more. Just as life forms are born in a niche and, if conditions allow, invade the entire ecosystem, similarly, if a galactic civilization overcame the seven challenges of the seventh parameter and reached the energy value of Kardashev **K = 1.4**, then it would invade the entire galaxy and become ETERNAL. In this case, the current civilizations would number over **2000**, all highly evolved and traveling in the galaxy.

The Fermi paradox, therefore, reappears in another form: it is not civilizations like ours that are missing and are residual but the super-civilizations hypothesized by Kardashev. The issue would be that we do not necessarily have to look for them on a specific planet because these civilizations also move between stars, and it would be much easier for them to find us rather than the other way around. So, the problem we should ask ourselves is: if we were a super-civilization capable of moving in the galaxy, where would we choose to go? Solar systems like ours (and we with it who, as we have seen, in addition

E. Mieli et al., *The Living Galaxy*, https://doi.org/10.1007/978-3-031-67324-5_23

to being alone as a **K1** civilization, we have a probability of only **0.4%** of becoming a **K2** civilization) may not be interesting for these civilizations. More likely, red dwarf stars (which last much longer than the sun) or even dead stars such as black holes and neutron stars (potentially exploitable as hypothetical energy sources) [110], would be more interesting.

The journey we have taken so far has led us to the following conclusions:

- **a** In the galaxy, as a primitive K1 civilization (actually less than K1), we are almost alone, with about 3 ETC including us currently present.
- **b** Approximately one civilization like ours forms every **20,000** years and has a slightly higher than **0.4%** chance of not going extinct (**1** in **250**).
- **c** There are nearly half a million civilizations like these already extinct in the galaxy.
- **d** Conversely, in the galaxy, if they overcame the seven challenges of the seventh parameter, there could be about **2000** super-civilizations, K2 level or almost, which would form one every **five million** years.
- **e** In this case, these super-civilizations would now be free to move between planetary systems and would likely be within about **50 light-years** from us (the distance of the first habitable planets used as intermediate travel stations). The organization and intentions of these super-civilizations are currently unknown to us and could be the subject of further study in our future work.
- **f** The rest of the galaxy is a jungle of life forms at various stages of development (tens of millions of planets inhabited by living forms).

While statement **d** and **e** on super-civilizations is based on the correct understanding of Drake's seventh parameter and therefore concerns futurology, all other statements, thanks to the mathematical treatment inspired by Maccone, are firmly anchored to their reference parameters and no longer range, as we have seen too many times, between tens and tens of orders of magnitude between the maximum and minimum values of uncertainty. Now a statement, for example, on the probability of the onset of prokaryotic life, must be strictly compared with the mathematical process we have reported: if we have to discuss or review it, we must also review the process that guides its parameter (in this case, the fourth parameter of Drake). This is the greatest advantage we have from this type of treatment and from the method inaugurated by Claudio Maccone in 2008 with his lognormal distribution.

Part V

Winners and Losers in the Milky Way

In this section, we will make simple assumptions about the fate of two possible intelligent alien species, each on their own planet (Fig. V.1). To achieve this, we will adhere only to the following constraints, dictated by common sense:

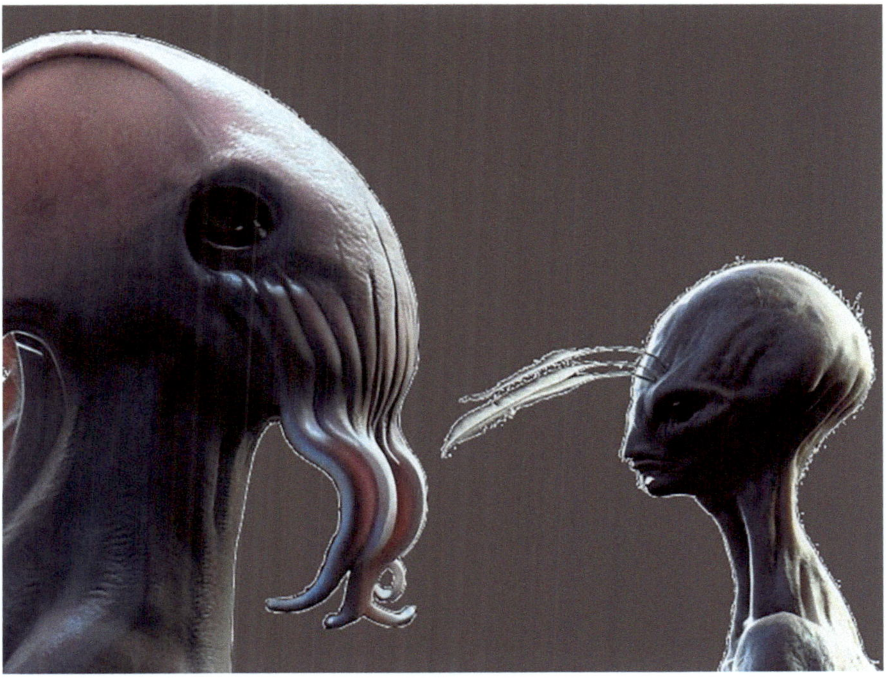

Fig. V.1 Winners and losers in the milky way: artistic representation of two galactic species represented by a cephalopoid and an arthropoid Powered by OpenAI

A Species have developed in systems similar to ours (**G** spectral class stars and planets in the habitable zone)

B On their respective planets, the development of phyla (taxonomic groups that rank below the animal kingdom) partially mirrors that on Earth.

C One of the two species is an arthropod (like insects), while the other is a mollusk (like cephalopods): we will call them the *arthropoid* and the *cephalopoid.*

D Apart from these similarities with Earth, the two species have evolved distinctly from their terrestrial *sister* species.

24

Identikit of Two Possible Intelligent Alien Species

Let's now look at the possible portrait of two species that have become intelligent according to the standards presented earlier. The first evolved from an arthropod-type anatomical plan, while the other evolved from the typical mollusk plan. If we have taken these paths, it is not for lack of imagination. This choice, that of using more or less known anatomical structures, can facilitate the reader's understanding. Moreover, who can say that these anatomical plans could not have led, under appropriate conditions, to the development of intelligent species?

24.1 The Arthropoid

The anatomical plan of the arthropod phylum provides for a body divided into various segments, each of which has a **pair of articulated limbs**. The main internal systems (digestive and nervous) are arranged asymmetrically to ours: in arthropods, the **nervous system is ventral** while the **digestive system is dorsal**. In vertebrates, the opposite occurs. Finally, the organism is equipped with an **external skeleton** (exoskeleton) to protect the soft tissues. This will be our starting point to build the first of our two intelligent creatures.

Since the rational capacity of our "arthropoid" is comparable to ours (and it reaches considerable body sizes, having a size that, more or less, is half of ours), some structural modifications are necessary. First of all, its nervous system has become as complex as ours, with a prominent brain rich in convolutions. The respiratory and circulatory systems have also evolved, to allow the species to acquire large dimensions. The circulatory system is closed (and not

E. Mieli et al., *The Living Galaxy*, https://doi.org/10.1007/978-3-031-67324-5_24

open as in current arthropods). The blood flows to the tissues and cells it needs to reach inside the arteries and returns from these to the heart through the veins. The heart is structured like ours: it consists of separate compartments where oxygen-rich blood does not mix with that poor in such gas. The heart is located in the chest, which is equipped with powerful lungs capable of pushing oxygenated blood throughout the body.

Let's describe our alien species from head to toe (Fig. 24.1):

- The skull is large, protected by the exoskeleton which includes the openings for the eyes (organs that have evolved to present the same characteristics as ours—quite different from those of current insects), for the masticatory apparatus (composed of a series of appendages modified in

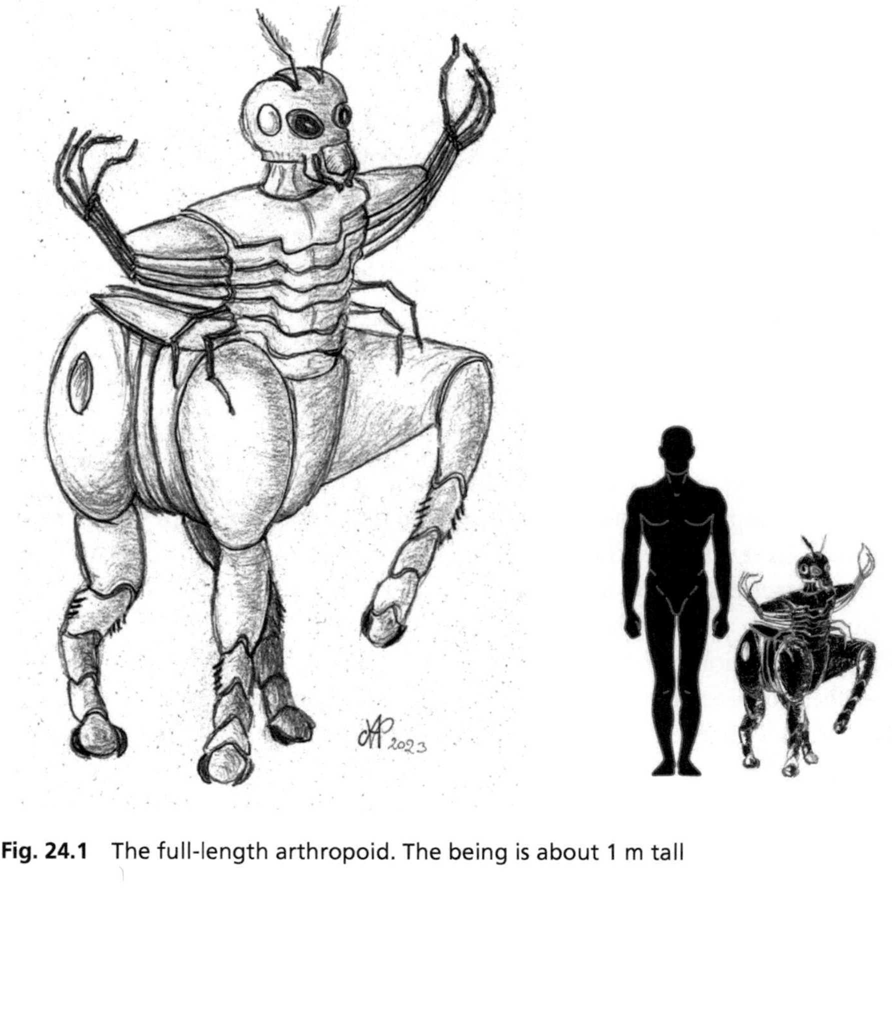

Fig. 24.1 The full-length arthropoid. The being is about 1 m tall

such a way as to support an omnivorous diet), and for the auditory apparatus (this species is capable of producing stridulations through a sound-producing organ located on the hip of its hind limbs). Finally, on the top of the skull, two antennae are arranged that allow the reception of chemical substances. These antennae can be erected or folded into special grooves behind their base, so as to be protected when they are not in use or when our creature needs to move at a fast speed.

- The thorax is composed of several elements, whose anatomy is relatively similar from element to element. The upper ones, under the neck, have articulated limbs with a high number of segments (the segments are the sections, articulated with each other, that constitute the limb in all its length). The proximal ones, closest to the thorax, have a wide base, are flattened and each can interlock with the equivalent of the lower (and upper) limb, in order to constitute a base capable of being subjected to a certain effort. The distal segments, furthest from the thorax, are instead completely independent of their homologues of the other elements, from the upper and the lower. The end of the limb corresponds to a "finger". Each finger, due to the independence mentioned above, is completely opposable to all the others on the same side of the body, allowing extremely effective manual dexterity.
- At the base of the thorax, two segments have thin limbs, capable of drumming a kind of Morse code on the exoskeleton. This system allows communication between individuals close to each other.
- Finally, in the upper part of the thorax, on the dorsal side, the respiratory slits open, allowing air to reach the lungs. Ventilation is ensured by an appropriate muscular system.
- The thorax is supported by the lower portion of the body, equipped with two pairs of limbs with highly developed musculature, in order to allow a sustained gait. The ventral part, well reinforced, protects the nervous system, while on the back the "shell" is less rigid, in order to allow the digestive system to fill properly when needed.
- In addition to the exoskeleton, rigid internal structures have evolved to allow the support of the musculature necessary for the functioning of a relatively large being.
- The exoskeleton requires regular molts, to allow the body to grow. In this species, body growth occurs in the juvenile phase; the size of the adult does not change anymore, as in mammals. Therefore, from "maturity", molts cease.

Our intelligent species has evolved from ***apo-diploid*** ancestors; that is, individuals with two copies of the genetic code turn out to be females, while those with only one copy are males (in the case of human beings, as in all vertebrates, all individuals possess two copies of DNA; we are all diploid). The apo-diploid system favors the formation of insect societies, as the workers, who are diploid (therefore females), are more closely related to their sisters (the other workers and queens with which they share **75%** of the genetic heritage instead of **50%**) and tend to help and favor their survival even at the expense of their offspring.

Our alien species still presents apo-diploidy, but, instead of building the large communities typical of most of our social insects, it forms smaller tribes. These, as happens, for example, in the current Argentine ant (*Linepithema humile*), are able to form larger colonies, living peacefully with other tribes.

The matriarchs reproduce and give birth to few larvae (usually two or three) at a very advanced stage of development. Later, they are able to mate again, so that the number of tribe members can increase relatively quickly. The matriarchs can mate with the males of their tribe or even with those of the tribes with which they coexist. This custom allows introducing "new blood" to the entire tribe and strengthening the bonds between the various groups of the entire colony.

Over the course of evolution, a particular form of "worker" has specialized: it has given rise to individuals specialized in reflection and intervention to improve the life of the tribe to which they belong. This is a caste that preludes the role of the scientist in intelligent society. The constitution of multi-tribal colonies leads to the emergence of individuals, "scientists", who operate not only for the good of their own tribe but also for the entire extended community. Deriving from workers of apo-diploid colonies, their purpose will be more collectivist than individualist. The impulse that moves them to improve the living conditions of the whole group leads them not only to favor the existence of individuals of the same species but also to the preservation of the environment and associated forms of life, understood as essential components of the life of the tribe and the entire colony.

At the beginning, the selection of individuals into various castes was established randomly and according to the mere needs of the community (for the next generation, "x" workers specialized in rearing larvae, "y" qualified for the well-being of the colony and "z" focused on its defense are needed). The larvae were therefore fed specific foods that, throughout their growth phase, developed the morphological and behavioral characteristics suitable for their future function. The morphological differences between the various castes were therefore very pronounced.

In the last evolutionary phase, however, all larvae are fed with rich regimes in the first phase of their growth. Only later, towards the end of adolescence, when individual characteristics allow recognizing individual propensities for various activities, does dietary differentiation intervene, capable of strengthening the physical characteristics proper to the specific function. The morphological differences are therefore much more reduced (naturally, the inequalities that characterize reproductive and non-reproductive individuals remain). In addition, the cultural transmission of the skills typical of each activity has increased.

We conclude with the geography occupied by the arthropoid. Their physiology has favored settlement in warm or temperate environments. However, the great architectural skills of our social insects do not make us doubt the structural realization capabilities that such forms of life can achieve. Quickly, our intelligent species has become capable of conceiving and realizing housing structures for the tribe (and for the entire colony) in which the conditions remain relatively constant, allowing all individuals to move without problems within them, regardless of external conditions.

Quickly, the technical abilities of our species have been able to produce coatings (practically clothes) to allow all adult individuals (remember that they no longer molt) to be able to withstand, being outside, various types of climate and therefore to colonize practically the entire dry surface from their home planet.

24.2 The Cephalopoid

The phylum of mollusks, our starting point for this new intelligent creature, is characterized by the possession of the **mantle**, a muscular tissue capable of secreting the shell (consisting of one or more variously mineralized elements), the **radula**, a kind of tongue equipped with "teeth" (very plastic; it allows processing a wide spectrum of foods) and the **foot**, muscular tissue from which derives the "sole" on which snails crawl or the arms of an octopus.

Our cephalopoid is endowed with a rational capacity comparable to ours and, to reach such an evolutionary level, it had to equip itself with a complex nervous system as well as circulatory and respiratory systems capable of performances similar to ours. In fact, the creature has a very high metabolism, with the need for significant amounts of oxygen and nutrients to function at full capacity.

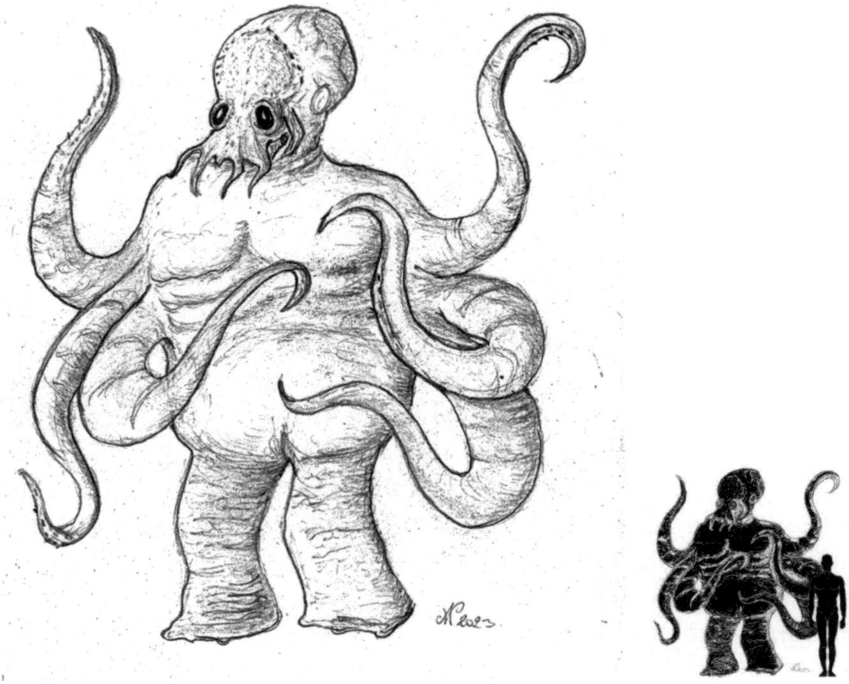

Fig. 24.2 The full-figure *cephalopoid.* The being is about 3 m tall

Let's now move on to a description of it from head to toe (Fig. 24.2):

- The being is equipped with a large skull that encloses a brain of impressive size. The eyes are like those of our modern octopuses. Between these organs and on the forehead there is a vast space rich in chromatophores, cells endowed with a large amount of pigments. The alternating lighting up and turning off and showing a particular color forms the basis of a complex "visual" language based on shapes and colors. It allows close communication between two or more individuals.
- The eyes are arranged to provide binocular vision. However, the neck, although robust, is very mobile and allows rotations of more than 100 degrees, to allow our intelligent creature to "watch its back".
- It does not have a nose, but it has a very developed sense of taste, the center of which is located inside the oral cavity (the mouth). The radula, complex, plays the role of teeth and allows the grinding and processing of various types of food, meat or vegetable. Above the mouth, or in the manner of

lips, are small modified tentacles, which resemble gills. Equipped with suckers and small hooks, they aid ingestion and a first processing of food.

- Still on the head, behind the eyes, are the modified gills (the distant ancestors of our creature inhabited aquatic environments) that allow the entry of air towards the lungs. The spiracle, behind the mouth, is a small tube that serves to expel carbon dioxide-rich air.
- The sense of hearing is relatively rudimentary and consists of small membranes located on the head, which can detect sound waves in a relatively limited frequency range.
- The brain and cephalic organs are protected by an internal shell, which serves as a skull for our creature.
- However, the cephalic shell is not the only mineralized internal support. Many others are distributed inside the body: they serve to support nerves and muscles and reinforce the body structure, as we will see later, in more detail.
- The chest is voluminous and contains a heart structured similarly to ours, in separate compartments, and efficient lungs capable of pushing blood throughout the body.
- From the chest, arms sprout (three per side) very similar superficially, to those of modern cephalopods. In reality, inside them there are mineralized reinforcements that make them, at least over a certain length in the proximal part (that is, the one closer to the body) behave similarly to the coils of snakes; they are therefore capable of clamping and crushing objects. In the distal part (the one towards the tip), however, such reinforcements are much more sparse. This allows this section of the arm to be partially extendable. Moreover, it is equipped with mini-suction cups and some hooks (although there is some individual variability in this character). In any case, the ends of such tentacles are rich in tactile papillae to exercise the sense of touch.
- Below the attachment of the arms, there is the abdomen, which houses the digestive system. Everything is supported by two strong limbs, similar to elephant legs, although they are more stubby and shorter than those of pachyderms. Our intelligent creature does not possess real nails, although, on the plantar surface that constitutes the sole of the feet, there are suction cups and small hooks, which aid in walking on slippery, inclined or irregular areas.
- However, the most important organs present on the lower surface of the legs are structures capable of detecting ground vibrations. Thanks to an adequate network of nerves, such sensations reach the brain where they are

processed. The individual can thus get a more or less precise idea of what is happening at a certain distance from him. He can therefore have a complete idea of the constitution of the environment, so as not to be caught off guard.

In any case, the size (3 m or more, from head to foot), the powerful musculature of the arms, the very robust hind legs, the abdomen and chest covered by a thick layer of fat (which gives tone to our creature and contributes to its thermoregulation since it is devoid of hair), as well as gregarious and social habits, contribute to the fact that our "hero" is practically devoid of natural enemies.

Regarding reproduction, the young develop in the body of the female. After birth, the young are cared for by both parents, but are quickly transferred to nurseries comprising other young of similar age. In such *nurseries*, the young learn the rudiments of their visual language and the basics for their everyday life. The social organization, initially based on the family system, has subsequently evolved to include a group composed of several families and/or single young individuals of both sexes.

Our alien species, endowed with acute vision, creates particularly sought-after dwellings for their aesthetics. Initially developed in a warm-temperate environment, thanks to the insulating external tissues of their body (initially evolved to avoid water loss) they quickly became capable of conquering various climatic zones of their planet. Technological acquisitions have made them masters of practically all the habitable land surfaces of their world.

24.3 The Arthropoid Faces the Seven Challenges of the Seventh Parameter

The arthropoid owes its evolutionary strength to the apo-diploid bond that it has extended not just to families of its own species, but effectively forming a global super-society. This allowed it to painlessly overcome the first challenge of the seventh parameter (the evolutionary inability to manage increasing amounts of energy): at the moment of transition to a civilization with **K = 0.6** to **K = 0.7** and beyond, it did not endanger itself and was able to implement the main countermeasures to prevent fatal errors (second challenge) (Fig. 24.3).

The first planetary changes of some significance (third challenge: in their case, for example, a supernova exploded within the safety distance) presented

Fig. 24.3 The city of the arthropoids in one of their *planet-colonies* Powered by OpenAI

themselves to the arthropoid when its Kardashev level had already reached beyond **K = 1**, and it already possessed the technologies and energy sufficient to face it. The problem of a possible spontaneous involution (fourth challenge) never arose because the division of the arthropoid society into castes always kept a part of the population on guard in defense of the *planetary hive*.

Challenges 5 and 6, on the other hand, were overcome with difficulty, since both risked shattering the orderly and granitic social structure of the arthropoids: genetic manipulation was therefore channeled so as not to damage the caste-divided social structure, and artificial intelligence was relegated to its own caste, without access to dangerous technological resources, but with the sole purpose of advising the guiding caste.

Today, the arthropoids face only the seventh challenge (**point Ω**), but they have now reached a value of **K = 1.8** and move around the galaxy as they wish, unconcerned about the fate of this or that particular planet.

THEY ARE AMONG THE WINNERS IN THE MILKY WAY

24.4 The Cephalopoid Faces the Seven Challenges of the Seventh Parameter

The cephalopoid bases its evolutionary success on becoming intelligent while being devoid of natural enemies on its planet. Its intelligence was mainly directed at adapting the planet's environment to the frequent changes to which it is subjected (the axis of rotation is unstable due to the absence of a heavy satellite like the moon; moreover, the tectonics of the cephalopoids' planet has little activity, so the carbon cycle struggles to stabilize the temperature). Its naturally peaceful nature allowed it to easily overcome the first and second challenges when its Kardashev level was around **K = 0.8** (ten times the energy we exploit on Earth). They had no problem with the third challenge since their species has long dedicated itself solely to solving environmental problems, all of which were resolved over the course of their existence (Fig. 24.4).

The fourth challenge put them in danger because they often felt they could lower their guard and regress. This did not happen thanks to the periodic climate changes that urged them not to become too distracted.

However, the fifth challenge (genetic transition) proved to be their downfall, as they inadvertently created a more resistant and aggressive version of themselves. This copy first exterminated them and then went extinct itself as a consequence of failing the first challenge: they self-destructed. The Kardashev level of the cephalopoids at the time of their extinction was **K = 1.1**, therefore quite high (**10,000** times the energy at our disposal).

THEY ARE AMONG THE LOSERS IN THE MILKY WAY

Fig. 24.4 The city of the cephalopoids during a period of planet overheating Powered by OpenAI

Part VI

Epilogue

25

The Cheetah's Pity

One final sociopolitical observation. A consequence of this analysis is that we have understood, at least in principle, where the real bottlenecks lie in the evolution of a galactic civilization:

A The first major filter is the forms of life that become ETCs (approximately **1 out of 70 planets hosting animal life**)
B A second dramatic reduction comes from the technological challenges of the seventh parameter (approximately **1 out of 250 ETCs that have developed**)

While we have no control over point **A**, point **B** tells us that the "great silence" resulting from the Fermi paradox warns us about the slim chances a civilization has to survive its own technological development. We should therefore understand that we will have to do everything possible to improve those ruthless odds of merely **0.4%** surviving our own technology. We must strive to raise those chances to less severe values that give us a minimum incentive to progress as a species and planet, despite everything.

So where do we stand in this *Universal Judgment* of Nature? The suspicion of not being entirely safe has been with us for a while, but perhaps we are not completely doomed, despite the ruthless survival rate against our own imperfections revealed in the last chapter. But let's take it one step at a time. What is the biggest immediate danger we must face? In fact, it is the first challenge of the seventh parameter—our instinctual insufficiency to manage increasing levels of available energy, which in simple terms means our radical inability to properly manage the planet as well as ourselves. But we mammals, what

E. Mieli et al., *The Living Galaxy*, https://doi.org/10.1007/978-3-031-67324-5_25

genetic tools (not cultural or ethical ones, which, as we see daily, can do very little at the level of natural awareness) do we have to address this gap? In fact, we do have one.

But let's take a step back. We know advantageous evolutionary traits most of the time do not emerge anew, but are modifications of some other trait intended for a completely different purpose. For example, human hands were not born to build tools, but for climbing. Is there then some inherent trait of mammals (and humans) that can provide our *natural awareness* of the future? Yes, there is: parental care—this behavior is the cornerstone on which mammals build their more or less complex and extended societies.

Other groups of living beings develop their sociality by following different paths (the super-genetic sharing of ants (Fig. 25.1), the exchange of symbiotic

Fig. 25.1 Social ants: as apodiploids, workers share **75%** of the genetic heritage and not **50%** Powered by OpenAI

bacteria in termites, etc.), but mammals predominantly use this parental care path which can be summarized as follows: the main goal is to preserve their offspring. Is that all? Not quite; male felines often kill the offspring of other males, hyenas often kill those they consider intruders in their clan, not to mention the monstrosities that humans have perpetrated against their own kind and the planet since their existence. It doesn't work trivially.

Let's move for a moment to the African savannah where a female cheetah has just isolated and captured a live antelope cub, her most common prey. The cub could not escape in any way, and so it does something unexpected: it tries to nurse from the cheetah's teat as if she were its mother. Incredibly, the female cheetah **lets it go** at this point, performing an action that appears to go against her own species' interests. The cub appealed for mercy, and the cheetah granted mercy—an interspecific act of mercy (Fig. 25.2).

Fig. 25.2 A female leopard together with an antelope cub isolated from the herd Powered by OpenAI

In general, purists of genetic selfishness tend to discard such behaviors as errors or evolutionary dead-ends, but they fail to account for a phenomenon in plain sight and well-described in game theory and decision theory (recall Nobel laureate John Forbes Nash Jr., of A Beautiful Mind)—namely, cooperation. Even cooperative behaviors have advantageous equilibrium points for the individual—in fact, more advantageous than those obtained from non-cooperative behaviors. Nature understands this well, and on this basis symbiosis, colonies of individuals, and cooperative societies exist even before their advantage becomes explicit. In other words, evolution prepares the groundwork for changes well before they manifest. Evolution, that is, prepares the groundwork for changes well before they occur.

But who actually safeguards this primary parental instinct among mammals, potentially extendable from their own offspring to the entire planet? Obviously the females, as if they had written it directly into their genetic heritage—not the males who at most make it coexist with other predominant behaviors, such as territoriality and sexual competition.

Then the answer to our problems is both terribly simple and terribly complex: males (men) SHOULD NOT govern or, at least, they should not decide the fate of communities because for them cooperation is always a choice among many other possible ones, despite the risk of self-destruction. For the female (woman), on the other hand, cooperation and the preservation of offspring (of life) is the only viable path because she is built for this purpose. Never as much as today does this primary female instinct, like the cheetah's pity, serve the planet (Fig. 25.3).

Fig. 25.3 Female of *Australopithecus africanus* (**3-2 My**) with her cub (**Jose Garcia and Renaud Joannes-Boyau/Southern Cross University**)

Acknowledgements

We thank Prof. **Stephen Webb** for the inspiration given on many of the topics covered and for the encouragement he has provided to complete this work. We also want to thank Dr. **Manuela Miggiano** for her useful suggestions on nucleic acids. Finally, we pay a due tribute to the philosopher **Hans Jonas** who, with his book "Gnosticism", gave us the cue we were looking for in our Introduction.

E. Mieli et al., *The Living Galaxy*, https://doi.org/10.1007/978-3-031-67324-5

Authors' Contributions

C. Maccone, with his main works *Mathematical SETI* and *EVO-SETI*, has indicated the method that has been followed throughout the work. He supervised the drafting and suggested valuable changes.

A. M. F. Valli took care of all the chemical, biological and paleontological part related to the fourth and fifth Drake parameter.

E. Mieli formulated the mathematical models used in all Drake parameters and developed the analysis related to the astronomical parameters (first, second and third) and the social ones (sixth and seventh).

E. Mieli et al., *The Living Galaxy*, https://doi.org/10.1007/978-3-031-67324-5

Figure Credits

Fig. 1.1 International Journal of Astrobiology, 14/06/2023—Mieli, Valli, Maccone
Fig. 1.2 International Journal of Astrobiology, 14/06/2023—Mieli, Valli, Maccone
Fig. 2.1 International Journal of Astrobiology, 14/06/2023—Mieli, Valli, Maccone
Fig. I.1 International Journal of Astrobiology, 14/06/2023—Mieli, Valli, Maccone
Fig. 3.1 Figure modified by Helen Klus, http://www.thestargarden.co.uk/Stars.html
Original image by Richard Powell/Penubag, https://commons.wikimedia.org/wiki/File:H-R_diagram_-edited-3.gif. Licensed under an Attribution-ShareAlike 4.0 International (CC BY-SA 4.0) license, https://creativecommons.org/licenses/by-sa/4.0/.
https://commons.wikimedia.org/wiki/File:HRDiagram_in_italian.gif
Fig. 4.1 NASA copyright policy states that "NASA material is not protected by copyright unless noted". (See Template:PD-USGov, NASA copyright policy page or JPL Image Use Policy). This file is in the public domain in the United States
Fig. 5.1 Figure made or composed by the authors
Fig. 5.2 Images made with artificial intelligence DELL-E 3
Fig. 5.3 European Southern Observatory (ESO)
https://www.eso.org/public/images/eso0828a/
Fig. 5.4 Images made with artificial intelligence DELL-E 3
Fig. 5.5 NASA, ESA and A. Schaller (for STScI)

E. Mieli et al., *The Living Galaxy*, https://doi.org/10.1007/978-3-031-67324-5

https://commons.wikimedia.org/wiki/File:Artist%27s_impression_of_an_ultra-short-period_planet.jpg#filelinks
Fig. 5.6 NASA's Goddard Space Flight Center Conceptual Image Lab
https://svs.gsfc.nasa.gov/20218#section_credits
Fig. 5.7 Figure made or composed by the authors
Fig. 5.8 Kevin Saff. The image is in the public domain
Fig. 5.9 ESA/NASA/SOHO/LASCO/EIT
https://www.esa.int/Science_Exploration/Space_Science/SOHO_overview2
Fig. 5.10 International Journal of Astrobiology, 14/06/2023—Mieli, Valli, Maccone
Fig. 6.1 International Journal of Astrobiology, 14/06/2023—Mieli, Valli, Maccone
Fig. II.1 Figure made or composed by the authors
Fig. 9.1 International Journal of Astrobiology, 14/06/2023—Mieli, Valli, Maccone
Fig. 10.1 Images made with artificial intelligence DELL-E 3
Fig. 10.2 International Journal of Astrobiology, 14/06/2023—Mieli, Valli, Maccone
Fig. 10.3 Ram Krishnamurthy—Center for Chemical Evolution—Scripps Research Institute
https://www.scripps.edu/krishnamurthy/The%20Prebiotic%20Depsipeptide%20Story.html
Fig. 10.4 International Journal of Astrobiology, 14/06/2023—Mieli, Valli, Maccone
Fig. 10.5 CNX OpenStax
https://cnx.org/contents/5CvTdmJL@4.4
Fig. 10.6 Images made with artificial intelligence DELL-E 3
Fig. 10.7 Images made with artificial intelligence DELL-E 3
Fig. 10.8 International Journal of Astrobiology, 14/06/2023—Mieli, Valli, Maccone
Fig. 10.9 Fvasconcellos 00:25, 19 April 2007 (UTC). The image is in the public domain
https://it.m.wikipedia.org/wiki/File:Adenosine_diphosphate_ribose_3D.png
Fig. 10.10 International Journal of Astrobiology, 14/06/2023—Mieli, Valli, Maccone
Fig. 10.11 Blausen.com staff (2014). "Medical gallery of Blausen Medical 2014". WikiJournal of Medicine 1 (2). DOI: https://doi.org/10.15347/wjm/2014.010. ISSN 2002-4436. The image is in the public domain
Fig. 10.12 Images made with artificial intelligence DELL-E 3
Fig. 10.13 Images made with artificial intelligence DELL-E 3

Fig. 10.14 Nastech
https://pubsapp.acs.org/cen/coverstory/84/8446cover.html
Fig. 10.15 International Journal of Astrobiology, 14/06/2023—Mieli, Valli, Maccone
Fig. 10.16 International Journal of Astrobiology, 14/06/2023—Mieli, Valli, Maccone
Fig. 10.17 International Journal of Astrobiology, 14/06/2023—Mieli, Valli, Maccone
Fig. 11.1 NASA—http://nai.arc.nasa.gov/library/images/news_articles/big_274_3.jpg
Fig. 13.1a fonte immagine
http://cnx.org/contents/GFy_h8cu@10.53:rZudN6XP@2/Introduction
https://commons.wikimedia.org/wiki/File:Figure_04_02_01.jpg
Fig. 13.1a (source: CNX OpenStax)
Fig. 13.1b Giac83
https://commons.wikimedia.org/wiki/File:Struttura_della_cellula_animale.svg
Fig. 13.2 International Journal of Astrobiology, 14/06/2023—Mieli, Valli, Maccone
Fig. 13.3 Images made with artificial intelligence DELL-E 3
Fig. 13.4 Images made with artificial intelligence DELL-E 3
Fig. 13.5 Christa Schleper/Nature journal
Fig. 13.6 International Journal of Astrobiology, 14/06/2023—Mieli, Valli, Maccone
Fig. 13.7 Centers for Disease Control and Prevention, part of the United States Department of Health and Human Services. The image is in the public domain.
Fig. 13.8 International Journal of Astrobiology, 14/06/2023—Mieli, Valli, Maccone
Fig. 13.9 Mariana Ruiz
https://commons.wikimedia.org/wiki/File:Cell_membrane_detailed_diagram_3.svg
Fig. 13.10 International Journal of Astrobiology, 14/06/2023—Mieli, Valli, Maccone
Fig. 13.11 International Journal of Astrobiology, 14/06/2023—Mieli, Valli, Maccone
Fig. 14.1 Jill Gregory / MOUNT SINAI HEALTH SYSTEM, licensed under CC-BY-ND
Fig. 14.2 International Journal of Astrobiology, 14/06/2023—Mieli, Valli, Maccone

Fig. 14.3 International Journal of Astrobiology, 14/06/2023—Mieli, Valli, Maccone
Fig. 14.4 Officer or employee of the United States Government as part of that person's official duties under the terms of Title 17, Chapter 1, Section 105 of the US Code. This work is in the public domain in the United States
Fig. 14.5 International Journal of Astrobiology, 14/06/2023—Mieli, Valli, Maccone
Fig. 15.1 Ryan Somma
https://commons.wikimedia.org/wiki/File:Life_in_the_Ediacaran_sea.jpg
Fig. 15.2 Eric Cheng/STANFOD UNIVERSITY
Fig. 15.3 Junnn11
https://commons.wikimedia.org/wiki/File:20191020_Yohoia_tenuis.png
Fig. 15.4 Figure made or composed by the authors
Fig. 15.5 International Journal of Astrobiology, 14/06/2023—Mieli, Valli, Maccone
Fig. 15.6 Misaki Ouchida/Gregory P. Wilson Mantilla /University of Washington
https://www.sci.news/paleontology/filikomys-primaevus-09014.html
Fig. 15.7 Figure made or composed by the authors
Fig. 15.8 sculture di Ron Seguin
http://www.idinosauri.it/curiosita3.html
Fig. 15.9 Images made with artificial intelligence DELL-E 3
Fig. 15.10 International Journal of Astrobiology, 14/06/2023—Mieli, Valli, Maccone
Fig. 15.11 Images made with artificial intelligence DELL-E 3
Fig. 15.12 International Journal of Astrobiology, 14/06/2023—Mieli, Valli, Maccone
Fig. 16.1 International Journal of Astrobiology, 14/06/2023—Mieli, Valli, Maccone
Fig. 18.1 International Journal of Astrobiology, 14/06/2023—Mieli, Valli, Maccone
Fig. 18.2 International Journal of Astrobiology, 14/06/2023—Mieli, Valli, Maccone
Fig. III.1 Images made with artificial intelligence DELL-E 3
Fig. 20.1 International Journal of Astrobiology, 14/06/2023—Mieli, Valli, Maccone
Fig. 20.2 International Journal of Astrobiology, 14/06/2023—Mieli, Valli, Maccone
Fig. 20.3 Images made with artificial intelligence DELL-E 3
Fig. 20.4 Images made with artificial intelligence DELL-E 3

Fig. 20.5 Images made with artificial intelligence DELL-E 3
Fig. 20.6 Images made with artificial intelligence DELL-E 3
Fig. 20.7 Images of the original paper
Fig. 20.8 Images of the original articles
https://www.dailygrail.com/2018/12/the-emperors-new-mind-roger-penrose-talks-to-joe-rogan-about-quantum-consciousness/
Fig. 20.9 Wgsimon
https://commons.wikimedia.org/wiki/File:Transistor_Count_and_Moore%27s_Law_-_2011.svg
Fig. 20.10 Images made with artificial intelligence DELL-E 3
Fig. 20.11 International Journal of Astrobiology, 14/06/2023—Mieli, Valli, Maccone
Fig. 20.12 International Journal of Astrobiology, 14/06/2023—Mieli, Valli, Maccone
Fig. IV.1a International Journal of Astrobiology, 14/06/2023—Mieli, Valli, Maccone
Fig. IV.1b International Journal of Astrobiology, 14/06/2023—Mieli, Valli, Maccone
Fig. V.1 Images made with artificial intelligence DELL-E 3
Fig. 24.1 Figure made or composed by the authors
Fig. 24.2 Figure made or composed by the authors
Fig. 24.3 Images made with artificial intelligence DELL-E 3
Fig. 24.4 Images made with artificial intelligence DELL-E 3
Fig. 25.1 Images made with artificial intelligence DELL-E 3
Fig. 25.2 Images made with artificial intelligence DELL-E 3
Fig. 25.3 Jose Garcia e Renaud Joannes-Boyau/Southern Cross University

Appendix A

Table 1 Summary of the 50 steps of the Drake equation

Step	Drake Par.	Value	Description
1	**DRAKE 1**	**$1.10 \times 10^{+10}$**	**NUMBER OF STARS IN THE GALAXY SUITABLE FOR LIFE (F, G AND K SPECTRAL CLASS)**
2	**DRAKE 2**	**1.80×10^{-1}**	**NUMBER OF SUITABLE PLANETS IN THE HABITABLE ZONE PER STAR (F, G AND K SPECTRAL CLASS)**
	DRAKE3	**1.84×10^{-2}**	**STABLE PLANETS FRACTION**
3			Multiple star systems
4			Supernovae at less than 40 ly
5			Gamma-ray bursts at less than 5,000 ly
6			Super flares of one's own star
7			Transit of gas giants on internal orbits
8			Prolonged meteorite bombardment
9			Instability of the rotation axis
10			Absence of the carbon cycle
11			Absence of the planetary magnetic field
	DRAKE 4	**5.16×10^{-1}**	**FRACTION OF PLANETS WHERE LIFE BORN**
12			The abiological synthesis of biological molecules
13			The concentration of the primordial soup
14			The formation of lipid pockets
15			The inclusion of chlorophyll in lipid membranes
16			The "proton photopump"
17			The formation of nucleic acid strands
18			The catalytic role of RNA
19			Determination of roles
20			Formation of the cell membrane
21			Emergence of the genetic code

E. Mieli et al., *The Living Galaxy*, https://doi.org/10.1007/978-3-031-67324-5

Step	Drake Par.	Value	Description
	DRAKE 5 eukaryotes	**$5,45 \times 10^{-1}$**	**FRACTION OF PLANETS WHERE EUKARYOTES ARE BORN**
22			The evolution of an aerobic bacterium
23			The host-symbiont encounter
24			The formation of pores and the escape of cytoplasmic extensions
25			The "envelopment" of the symbionts and the disappearance of the host's cell wall
26			The penetration of the symbionts into the cytoplasm
27			The migration of DNA from the symbiont genome to that of the host
28			The acquisition of the eukaryotic cytoplasmic membrane
29			Incorporation into a single coat and phagocytosis
	DRAKE 5 metazoans	**8.49×10^{-1}**	**FRACTION OF PLANETS WHERE ANIMALS (METAZOA) ARE BORN**
30			The acquisition of a complex life cycle
31			The aggregation of zoospores and the formation of synzoospores
32			The sedentary colony composed of differentiated cells
33			The production of collagen
	DRAKE 5 ETC	**3.48×10^{-2}**	**FRACTION OF PLANETS WHERE TECHNOLOGICAL CIVILIZATIONS ARE BORN (ETC)**
34			Increase in metazoan size (nervous and vascular system)
35			Arts development
36			Conquers mainland
37			Differentiation of terrestrial animals
38			Acquisition of sociability
39			Standing position and manual dexterity
40			Change in diet and brain growth
41			Organization of the brain on abstract thought
42			Birth of articulated language and technique
43	**DRAKE 6**	**5.00×10^{-1}**	**FRACTION OF PLANETS WHERE LIFE DECIDES TO COMMUNICATE**
	DRAKE 7	**7.29×10^{-6}**	**STATIC ETC DURATION FRACTION ($K < 1.4$)**
44			Self-destruction due to evolutionary failure
45			Inadvertent technological error
46			Technological insufficiency to deal with planetary changes
47			Spontaneous involution
48			Artificial genetic transition ended on a dead end
49			Artificial intelligence transition ended on a dead end
50			Reaching point Ω
	DRAKE 7	**4.64×10^{-3}**	**FRACTION OF DYNAMIC ETCs THAT OVERCOME THE 7 CHALLENGES AND BECOME ETERNAL ($K \geq 1.4$)**

Appendix B: Percentage of Past Planets and Present for Each Stage of Development

The calculation sequence reported in these tables starts from Drake's first three astronomical parameters, providing their probabilistic combination for the entire duration of the planet, from **0** to **9.5 billion** years (**Gy**).

The subsequent combination with the following terms, from the fourth parameter onwards, takes into account the necessary timescales when the processes reported in Table IV.3 (Part IV) occur. Based on those timescales, the corresponding value of the third parameter is chosen.

Up until the sixth parameter, both the number of planets inhabited by life forms during the entire **7 Gy** stellar population lifespan, and the number of those inhabited planets, are reported. The seventh parameter instead introduces a rigid temporal constraint that directly provides the current number of extant technological civilizations (ETCs), both in the case of static K1 civilizations (**7A**) and dynamic K2 civilizations (**7B**).

Finally, we note that the distance values reported in Table IV.4 (Part IV) are obtained taking into account the following average values of typical quantities in the galactic disk, that is, without considering the galactic center (Bulge):

$\mathbf{1.13 \times 10^{11}}$	Average number of disk stars
$\mathbf{1.53 \times 10^{13}}$	Average disk volume (**ly**^{3})
5	Average distance stars (**ly**)

E. Mieli et al., *The Living Galaxy*, https://doi.org/10.1007/978-3-031-67324-5

Appendix B1

Table 2 Percentage of past planets and present for each stage of development

1°-2°-3° Drake	grandezza sperimentale		frazione minima nel tempo max	frazione massima nel tempo max	componente j media logaritmo	componente j varianza logaritmo	somma media logaritmo	somma varianza logaritmo	media processo tot	scostamento processo tot	MIN pianeti potenzoalmente abitabili	MAX pianeti potenzoalmente abitabili		
		ΔT	A_j	B_j	μ_j	σ^2_j	μ	σ^2	$\langle X_0\rangle$	σX_0	$\langle X0\rangle_A$	$\langle X0\rangle_B$		
1	Numero di stelle della galassia adatte alla vita (di classe spettrale F, G e K)	anni di durata	1,00E+10	1,20E+10	23,1198	0,0028								
2	numero pianeti adatti nella zona abitabile per stella (di classe spettrale F, G e K)	del pianeta	1,60E-01	2,00E-01	-1,7169	0,0041								
3	frazione di pianeti stabili per anni	0,00E+00	1,00E+00	1,00E+00	0,0000	0,0000	21,4029	0,0069	1,98E+09	1,65E+08	1,69E+09	2,27E+09	5,E+08 passo temporale	
		5,00E+08	6,92E-01	7,80E-01	-0,3067	0,0012	21,0962	0,0081	1,46E+09	1,31E+08	1,23E+09	1,69E+09	τ MIN = 1.359.230.076	anni
		1,00E+09	4,79E-01	6,09E-01	-0,6111	0,0048	20,7918	0,0117	1,08E+09	1,17E+08	8,75E+08	1,28E+09	τMAX = 2.016.147.681	anni
		1,50E+09	3,32E-01	4,75E-01	-0,9130	0,0107	20,4899	0,0176	7,99E+08	1,06E+08	6,14E+08	9,83E+08	p = exp (-ΔTMAX/τ)	
		2,00E+09	2,30E-01	3,71E-01	-1,2126	0,0189	20,1903	0,0258	5,94E+08	9,62E+07	4,28E+08	7,61E+08		
		2,50E+09	1,59E-01	2,89E-01	-1,5099	0,0294	19,8930	0,0363	4,44E+08	8,54E+07	2,96E+08	5,92E+08		
		3,00E+09	1,10E-01	2,26E-01	-1,8048	0,0420	19,5981	0,0489	3,33E+08	7,45E+07	2,04E+08	4,62E+08		
		3,50E+09	7,62E-02	1,76E-01	-2,0975	0,0567	19,3054	0,0636	2,50E+08	6,41E+07	1,39E+08	3,61E+08		
		4,00E+09	5,27E-02	1,38E-01	-2,3879	0,0732	19,0150	0,0801	1,89E+08	5,45E+07	9,42E+07	2,83E+08		
		4,50E+09	3,65E-02	1,07E-01	-2,6762	0,0916	18,7267	0,0985	1,43E+08	4,59E+07	6,32E+07	2,22E+08		
		5,00E+09	2,53E-02	8,37E-02	-2,9623	0,1116	18,4406	0,1185	1,08E+08	3,84E+07	4,17E+07	1,75E+08		
		5,50E+09	1,75E-02	6,54E-02	-3,2464	0,1331	18,1566	0,1400	8,24E+07	3,19E+07	2,71E+07	1,38E+08		
		6,00E+09	1,21E-02	5,10E-02	-3,5284	0,1559	17,8745	0,1628	6,28E+07	2,64E+07	1,71E+07	1,09E+08		
		6,50E+09	8,38E-03	3,98E-02	-3,8085	0,1799	17,5945	0,1868	4,81E+07	2,18E+07	1,03E+07	8,58E+07		
		7,00E+09	5,80E-03	3,11E-02	-4,0867	0,2050	17,3163	0,2119	3,68E+07	1,79E+07	5,84E+06	6,78E+07		
		7,50E+09	4,01E-03	2,42E-02	-4,3630	0,2308	17,0399	0,2377	2,83E+07	1,47E+07	2,91E+06	5,37E+07		
		8,00E+09	2,78E-03	1,89E-02	-4,6376	0,2574	16,7653	0,2643	2,18E+07	1,20E+07	1,03E+06	4,26E+07		
		8,50E+09	1,92E-03	1,48E-02	-4,9106	0,2845	16,4923	0,2914	1,68E+07	9,78E+06	-1,24E+05	3,38E+07		
		9,00E+09	1,33E-03	1,15E-02	-5,1819	0,3119	16,2210	0,3188	1,30E+07	7,97E+06	-7,98E+05	2,68E+07		
		9,50E+09	9,22E-04	8,99E-03	-5,4517	0,3396	15,9512	0,3465	1,01E+07	6,48E+06	-1,15E+06	2,13E+07		

4° Drake	grandezza sperimentale		frazione minima nel tempo max	frazione massima nel tempo max	componente j media logaritmo	componente j varianza logaritmo	somma media logaritmo	somma varianza logaritmo	media processo tot	scostamento processo tot	MIN processo tot	MAX processo tot		
		ΔT	A_j	B_j	μ_j	σ^2_j	μ	σ^2	$\langle X_0\rangle$	σX_0	$\langle X0\rangle_A$	$\langle X0\rangle_B$		
1	Numero di stelle della galassia adatte alla vita (di classe spettrale F, G e K)	anni di durata	1,00E+10	1,20E+10	23,1198	0,0028	20,3603	0,0490	7,13E+08	1,60E+08	4,36E+08	9,89E+08	712.851.649	91.652.355
2	numero pianeti adatti nella zona abitabile per stella (di classe spettrale F, G e K)	del pianeta	1,60E-01	2,00E-01	-1,7169	0,0041							pianeti in 7 Ga su un totale di 1.457.862.814	pianeti attuali su un totale di 187.439.505
3	frazione di pianeti stabili per anni	9,00E+08	6,92E-01	7,80E-01	-0,3067	0,0012								
4	frazione di pianeti dove nasce la vita		3,22E-01	6,55E-01	-0,7359	0,0409								

5° eucarioti	grandezza sperimentale		frazione minima nel tempo max	frazione massima nel tempo max	componente j media logaritmo	componente j varianza logaritmo	somma media logaritmo	somma varianza logaritmo	media processo tot	scostamento processo tot	MIN processo tot	MAX processo tot		
		ΔT	A_j	B_j	μ_j	σ^2_j	μ	σ^2	$\langle X_0\rangle$	σX_0	$\langle X0\rangle_A$	$\langle X0\rangle_B$		
1	Numero di stelle della galassia adatte alla vita (di classe spettrale F, G e K)	anni di durata	1,00E+10	1,20E+10	23,1198	0,0028	18,1358	0,1541	8,12E+07	3,32E+07	2,38E+07	1,39E+08	81.238.609	34.816.547
2	numero pianeti adatti nella zona abitabile per stella (di classe spettrale F, G e K)	del pianeta	1,60E-01	2,00E-01	-1,7169	0,0041							pianeti in 7 Ga su un totale di 332.630.617	pianeti attuali su un totale di 142.555.979
3	frazione di pianeti stabili per anni	3,00E+09	1,10E-01	2,26E-01	-1,8048	0,0420								
4	frazione di pianeti dove nasce la vita		3,22E-01	6,55E-01	-0,7359	0,0409								
5 eucarioti	frazione di pianeti dove nascono eucarioti		2,89E-01	7,09E-01	-0,7264	0,0644								

Appendix B2

Table 3 Percentage of past planets and present for each stage of development

5° metazoi	grandezza sperimentale		frazione minima nel tempo max	frazione massima nel tempo max	componente j media logaritmo	componente j varianza logaritmo	somma media logaritmo	somma varianza logaritmo	media processo tot	scostamento processo tot	MIN processo tot	MAX processo tot		
		ΔT	A_j	B_j	μ_j	σ^2_j	μ	σ^2	$\langle X_0\rangle$	σX_0	$\langle X0\rangle_A$	$\langle X0\rangle_B$		
1	Numero di stelle della galassia adatte alla vita (di classe spettrale F, G e K)	anni di durata	1,00E+10	1,20E+10	23,1198	0,0028	17,2836	0,2026	3,55E+07	1,68E+07	6,36E+06	6,46E+07	35.493.456	20.281.975
2	numero pianeti adatti nella zona abitabile per stella (di classe spettrale F, G e K)	del pianeta	1,60E-01	2,00E-01	-1,7169	0,0041							pianeti in 7 Ga su un totale di 188.580.549	pianeti attuali su un totale di 107.760.314
3	frazione di pianeti stabili per anni	4,00E+09	5,27E-02	1,38E-01	-2,3879	0,0732								
4	frazione di pianeti dove nasce la vita		3,22E-01	6,55E-01	-0,7359	0,0409								
5 eucarioti	frazione di pianeti dove nascono eucarioti		2,89E-01	7,09E-01	-0,7264	0,0644								
5 metazoi	frazione di pianeti dove nascono metazoi		5,97E-01	9,44E-01	-0,2692	0,0172								

5° CET	grandezza sperimentale		frazione minima nel tempo max	frazione massima nel tempo max	componente j media logaritmo	componente j varianza logaritmo	somma media logaritmo	somma varianza logaritmo	media processo tot	scostamento processo tot	MIN processo tot	MAX processo tot		
		ΔT	A_j	B_j	μ_j	σ^2_j	μ	σ^2	$\langle X_0\rangle$	σX_0	$\langle X0\rangle_A$	$\langle X0\rangle_B$		
1	Numero di stelle della galassia adatte alla vita (di classe spettrale F, G e K)	anni di durata	1,00E+10	1,20E+10	23,1198	0,0028	13,5528	0,3909	9,35E+05	6,47E+05	-1,85E+05	2,06E+06	934.994	601.067
2	numero pianeti adatti nella zona abitabile per stella (di classe spettrale F, G e K)	del pianeta	1,60E-01	2,00E-01	-1,7169	0,0041							pianeti in 7 Ga su un totale di 142.657.715	pianeti attuali su un totale di 91.708.531
3	frazione di pianeti stabili per anni	4,50E+09	3,65E-02	1,07E-01	-2,6762	0,0916								
4	frazione di pianeti dove nasce la vita		3,22E-01	6,55E-01	-0,7359	0,0409								
5 eucarioti	frazione di pianeti dove nascono eucarioti		2,89E-01	7,09E-01	-0,7264	0,0644								
5 metazoi	frazione di pianeti dove nascono metazoi		5,97E-01	9,44E-01	-0,2692	0,0172								
5 CET	frazione di pianeti dove nascono CET		1,25E-02	5,66E-02	-3,4425	0,1700								

6° Drake	grandezza sperimentale		frazione minima nel tempo max	frazione massima nel tempo max	componente j media logaritmo	componente j varianza logaritmo	somma media logaritmo	somma varianza logaritmo	media processo tot	scostamento processo tot	MIN processo tot	MAX processo tot		
		ΔT	A_j	B_j	μ_j	σ^2_j	μ	σ^2	$\langle X_0\rangle$	σX_0	$\langle X0\rangle_A$	$\langle X0\rangle_B$		
1	Numero di stelle della galassia adatte alla vita (di classe spettrale F, G e K)	anni di durata	1,00E+10	1,20E+10	23,1198	0,0028	12,8529	0,4045	4,68E+05	3,30E+05	-1,04E+05	1,04E+06	467.518	300.547
2	numero pianeti adatti nella zona abitabile per stella (di classe spettrale F, G e K)	del pianeta	1,60E-01	2,00E-01	-1,7169	0,0041							pianeti in 7 Ga su un totale di 142.657.715	pianeti attuali su un totale di 91.708.531
3	frazione di pianeti stabili per anni	4,50E+09	3,65E-02	1,07E-01	-2,6762	0,0916								
4	frazione di pianeti dove nasce la vita		3,22E-01	6,55E-01	-0,7359	0,0409								
5 eucarioti	frazione di pianeti dove nascono eucarioti		2,89E-01	7,09E-01	-0,7264	0,0644								
5 metazoi	frazione di pianeti dove nascono metazoi		5,97E-01	9,44E-01	-0,2692	0,0172								
5 CET	frazione di pianeti dove nascono CET		1,25E-02	5,66E-02	-3,4425	0,1700								
6	frazione di pianeti dove la vita decide di comunicare		4,00E-01	6,00E-01	-0,6999	0,0136								

Appendix B3

Table 4 Percentage of past planets and present for each stage of development

7° A Drake	grandezza sperimentale		frazione minima nel tempo max	frazione massima nel tempo max	componente j media logaritmo	componente j varianza logaritmo	somma media logaritmo	somma varianza logaritmo	media processo tot	scostamento processo tot	MIN processo tot	MAX processo tot	
		ΔT	A_j	B_j	μ_j	σ^2_j	μ	σ^2	$\langle X_0\rangle$	σX_0	$\langle X0\rangle_A$	$\langle X0\rangle_B$	
1	Numero di stelle della galassia adatte alla vita (di classe spettrale F, G e K)	anni di durata	1,00E+10	1,20E+10	23,1198	0,0028	1,0213	0,4110	3,41E+00	2,43E+00	-8,01E-01	7,62E+00	3
2	numero pianeti adatti nella zona abitabile per stella (di classe spettrale F, G e K)	del pianeta	1,60E-01	2,00E-01	-1,7169	0,0041							pianeti attuali su un totale di 142.657.715
3	frazione di pianeti stabili per anni	4,55E+09	3,65E-02	1,07E-01	-2,6762	0,0916							
4	frazione di pianeti dove nasce la vita		3,22E-01	6,55E-01	-0,7359	0,0409							
5 eucarioti	frazione di pianeti dove nascono eucarioti		2,89E-01	7,09E-01	-0,7264	0,0644							
5 metazoi	frazione di pianeti dove nascono metazoi		5,97E-01	9,44E-01	-0,2692	0,0172							
5 CET	frazione di pianeti dove nascono CET		1,25E-02	5,66E-02	-3,4425	0,1700							
6	frazione di pianeti dove la vita decide di comunicare		4,00E-01	6,00E-01	-0,6999	0,0136							
7	frazione temporale della durata di una civiltà di tipo K1 statica		6,29E-06	8,30E-06	-11,8316	0,0064							

7° B Drake	grandezza sperimentale		frazione minima nel tempo max	frazione massima nel tempo max	componente j media logaritmo	componente j varianza logaritmo	somma media logaritmo	somma varianza logaritmo	media processo tot	scostamento processo tot	MIN processo tot	MAX processo tot	
		ΔT	A_j	B_j	μ_j	σ^2_j	μ	σ^2	$\langle X_0\rangle$	σX_0	$\langle X0\rangle_A$	$\langle X0\rangle_B$	
1	Numero di stelle della galassia adatte alla vita (di classe spettrale F, G e K)	anni di durata	1,00E+10	1,20E+10	23,1198	0,0028	7,3533	0,6981	2,21E+03	2,22E+03	-1,64E+03	6,07E+03	2.214
2	numero pianeti adatti nella zona abitabile per stella (di classe spettrale F, G e K)	del pianeta	1,60E-01	2,00E-01	-1,7169	0,0041							pianeti attuali su un totale di 142.657.715
3	frazione di pianeti stabili per anni	4,95E+09	3,65E-02	1,07E-01	-2,6762	0,0916							
4	frazione di pianeti dove nasce la vita		3,22E-01	6,55E-01	-0,7359	0,0409							
5 eucarioti	frazione di pianeti dove nascono eucarioti		2,89E-01	7,09E-01	-0,7264	0,0644							
5 metazoi	frazione di pianeti dove nascono metazoi		5,97E-01	9,44E-01	-0,2692	0,0172							
5 CET	frazione di pianeti dove nascono CET		1,25E-02	5,66E-02	-3,4425	0,1700							
6	frazione di pianeti dove la vita decide di comunicare		4,00E-01	6,00E-01	-0,6999	0,0136							
7	frazione temporale della durata di una civiltà di tipo K2 dinamica		1,03E-03	8,25E-03	-5,4996	0,2935							

Appendix C: The Calculation of the Distribution Function of the Seventh Parameter

Knowing that $\boldsymbol{\Phi}(z)$ is Maccone's lognormal, the $F_p(\Delta T_p)$ distribution function is written:

$$
\begin{cases}
\Phi(\mathrm{z}) \equiv \dfrac{1}{\mathrm{z}} \dfrac{1}{\sqrt{2\pi\sigma}} e^{-\frac{(\ln(\mathrm{z})-\mu)^2}{2\sigma^2}} \\[2ex]
\mathbf{F_p}(\Delta \mathbf{T_p}) = \Phi\left(p^{\frac{\Delta T_0}{\Delta T_p}} \right) p^{\frac{\Delta T_0}{\Delta T_p}} \left(\dfrac{\Delta T_0 \ln \frac{1}{p}}{\Delta T_p{}^2} \right)
\end{cases}
$$

E. Mieli et al., *The Living Galaxy*, https://doi.org/10.1007/978-3-031-67324-5

To derive the above equation with respect to $\boldsymbol{\Delta T_p}$, we use the following formulas:

$$\begin{cases} \dfrac{d}{dz}\Phi(z)=\Phi(z)\cdot\left[\dfrac{\mu-\ln(z)-\sigma^2}{z\sigma^2}\right] \\ \dfrac{d}{dy}\left(\boldsymbol{p}^{\frac{y_0}{y}}\right)=\dfrac{y_0\cdot\ln\left(\dfrac{1}{p}\right)\cdot\boldsymbol{p}^{\frac{y_0}{y}}}{y^2} \\ \dfrac{d}{dy}\left[\boldsymbol{p}^{\frac{y_0}{y}}\cdot\left(\dfrac{y_0\cdot\ln\dfrac{1}{\boldsymbol{p}}}{y^2}\right)\right]=\dfrac{y_0\cdot\ln\left(\dfrac{1}{p}\right)\cdot\boldsymbol{p}^{\frac{y_0}{y}}}{y^4}y_0\cdot\ln\left(\dfrac{1}{p}\right)-2y \end{cases}$$

where we have set, as a synthetic notation: $\boldsymbol{\Delta T_p = y}$ and $\boldsymbol{\Delta T_0 = y_0}$.

Therefore, we will have:

$$\frac{d}{dy}F(y)=\Phi\left(\boldsymbol{p}^{\frac{y_0}{y}}\right)\cdot\frac{y_0\cdot\ln\left(\dfrac{1}{p}\right)\cdot\boldsymbol{p}^{\frac{y_0}{y}}}{y^5\sigma^2}\cdot\left[\begin{array}{l}\left(-2\sigma^2\right)\cdot y^2+\left(\mu\cdot y_0\cdot\ln\left(\dfrac{1}{p}\right)\right)\cdot y\\ +\left(y_0\cdot\ln\left(\dfrac{1}{p}\right)\right)^2\end{array}\right]=0$$

That is to say:

$$\left(-2\sigma^2\right)\cdot y^2+\left(\mu\cdot y_0\cdot\ln\left(\frac{1}{p}\right)\right)\cdot y+\left(y_0\cdot\ln\left(\frac{1}{p}\right)\right)^2=0$$

and then, selecting only the greater-than-zero solution for $\boldsymbol{y}$:

$$\begin{cases} y_{MAX}=\mathbf{C}\left(y_0\cdot\ln\left(\dfrac{1}{p}\right)\right) \\ \mathbf{C}=\dfrac{\mu+\sqrt{\mu^2+8\sigma^2}}{4\sigma^2} \end{cases}$$

Finally, at the maximum point, for $\boldsymbol{F_p}(\boldsymbol{\Delta T_p})$, you will have:

$$\begin{cases} F\left(\Delta T_{pMAX}\right) = \boldsymbol{D}\dfrac{1}{y_0 \cdot \ln\left(\dfrac{1}{p}\right)} \\ \mathbf{D} = \dfrac{\exp\left[-\dfrac{1}{2\sigma^2}\left(\dfrac{4\sigma^2}{\sqrt{\mu^2+8\sigma^2}+\mu}+\mu\right)^2\right]}{\sqrt{2\pi}\sigma\left(\dfrac{\sqrt{\mu^2+8\sigma^2}+\mu}{4\sigma^2}\right)^2} \end{cases}$$

Appendix D: The Complete Calculation of the Lognormal

Step 1: Letting each factor become a random variable

In this paper, we adopt the notations of the great book "Probability, Random Variables and Stochastic Processes" by Athanasios Papoulis (1921–2002), now re-published as Papoulis-Pillai. The advantage of this notation is that it makes a neat distinction between probabilistic (or statistical: it is the same thing here) variables, always denoted by **capitals**, from non-probabilistic (or "deterministic") variables, always denoted by **lower-case** letters. Adopting the Papoulis notation also is a tribute to him by this author, who was a Fulbright Grantee in the United States with him at the Polytechnic Institute (now Polytechnic University) of New York in the years 1977–79.

We thus introduce seven new (positive) random variables Di ("D" from "Drake") defined as:

$$\begin{cases} D_1 = Ns \\ D_2 = fp \\ D_3 = ne \\ D_4 = fl \\ D_5 = fi \\ D_6 = fc \\ D_7 = fL \end{cases} \tag{2}$$

E. Mieli et al., *The Living Galaxy*, https://doi.org/10.1007/978-3-031-67324-5

so that our ***STATISTICAL Drake equation*** may be simply rewritten as

$$N = \prod_{i=1}^{7} D_i. \tag{3}$$

Of course, N now becomes a (positive) random variable too, having its own (positive) mean value and standard deviation. Just as each of the Di has its own (positive) mean value and standard deviation…

… the natural question then arises: how are the seven mean values on the right related to the mean value on the left?

… and how are the seven standard deviations on the right related to the standard deviation on the left?

Just take the next step, STEP TWO.

2.1. Step 2: Introducing logs to change the product into a sum

Products of random variables are not easy to handle in probability theory. It is actually much easier to handle sums of random variables, rather than products, because:

1. The probability density of the sum of two or more independent random variables is the convolution of the relevant probability densities (worry not about the equations, right now).
2. The Fourier transform of the convolution simply is the product of the Fourier transforms (again, worry not about the equations, at this point).

So, let us take the natural logs of both sides of the Statistical Drake Eq. (3) and change it into a sum:

$$\ln(N) = \ln\left(\prod_{i=1}^{7} D_i\right) = \sum_{i=1}^{7} \ln(D_i). \tag{4}$$

It is now convenient to introduce eight new (positive) random variables defined as follows:

$$\begin{cases} Y = \ln(N) \\ Y_i = \ln(D_i) \quad i = 1, \ldots, 7. \end{cases} \tag{5}$$

Upon inversion, the first equation of Eq. (5) yields the important equation, that will be used in the sequel

$$N = e^{Y}. \tag{6}$$

We are now ready to take STEP THREE.

2.2. Step 3: The transformation law of random variables

So far we did not mention at all the problem: "which probability distribution shall we attach to each of the seven (positive) random variables Di?"

It is not easy to answer this question because we do not have the least scientific clue to what probability distributions fit at best to each of the seven points listed in Section 1.

Yet, at least one trivial error must be avoided: claiming that each of those seven random variables must have a Gaussian (i.e. normal) distribution. In fact, the Gaussian distribution, having the well-known bell-shaped probability density function

$$f_X(x;\mu,\sigma) = \frac{1}{\sqrt{2\pi}\sigma} e^{-\frac{(x-\mu)^2}{2\sigma^2}} \quad (\sigma \geq 0) \tag{7}$$

has its independent variable x ranging between $-\infty$ and ∞ and so it can apply to a real random variable X only, and never to positive random variables like those in the statistical Drake Eq. (3). Period.

Searching again for probability density functions that represent positive random variables, an obvious choice would be the gamma distributions. However, we discarded this choice too because of a different reason: please keep in mind that, according to Eq. (5), once we selected a particular type of probability density function (pdf) for the last seven of Eq. (5), then we must compute the (new and different) pdf of the logs of such random variables. And the pdf of these logs certainly is not gamma-type any more.

It is high time now to remind the reader of a certain theorem that is proved in probability courses, but, unfortunately, does not seem to have a specific name. It is the transformation law (so we shall call it, see, for instance, Ref. [5]) allowing us to compute the pdf of a certain new random variable Y that is a known function Y = g(X) of an another random variable X having a known pdf. In other words, if the pdf fX(x) of a certain random variable X is known, then the pdf fY(y) of the new random variable Y, related to X by the functional relationship

$$Y = g(X) \tag{8}$$

can be calculated according to this rule:

1. First, invert the corresponding non-probabilistic equation y = g(x) and denote by xi(y) the various real roots resulting from this inversion.

2. Second, take notice whether these real roots may be either finitely- or infinitely many, according to the nature of the function y = g(x).
3. Third, the probability density function of Y is then given by the (finite or infinite) sum

$$f_Y(y) = \sum_i \frac{f_X(x_i(y))}{\left|g'(x_i(y))\right|} \tag{9}$$

where the summation extends to all roots xi (y) and |g′(xi (y))| is the absolute value of the first derivative of g(x), where the i-th root xi (y) has been replaced instead of x.

Since we must use this transformation law to transfer from the Di to the Yi = ln (Di), it is clear that we need to start from a Di pdf that is as simple as possible. The gamma pdf is not responding to this need because the analytic expression of the transformed pdf is very complicated (or, at least, it looked so to this author in the first instance). Also, the gamma distribution has two free parameters in it, and this "complicates" its application to the various meanings of the Drake equation. In conclusion, we discarded the gamma distributions and confined ourselves to the simpler uniform distribution instead, as shown in the next section.

3. Step 4: Assuming the easiest input distribution for each Di: the uniform distribution

Let us now suppose that each of the seven Di is distributed UNIFORMLY in the interval ranging from the lower limit ai ≥ 0 to the upper limit bi ≥ ai.

This is the same as saying that the probability density function of each of the seven Drake random variables D_i has the equation

$$f_{\text{uniform}\_D_i}(x) = \frac{1}{b_i - a_i} \quad \text{with} \quad 0 \le a_i \le x \le b_i \tag{10}$$

as it follows at once from the normalization condition

$$\int_{a_i}^{b_i} f_{\text{uniform}\_D_i}(x)\,dx = 1. \tag{11a}$$

Let us now consider the mean value of such uniform Di defined by

$$\begin{aligned}\langle \text{uniform}\_D_i \rangle &= \int_{a_i}^{b_i} x\, f_{\text{uniform}\_D_i}(x)dx = \frac{1}{b_i - a_i}\int_{a_i}^{b_i} x dx \\ &= \frac{1}{b_i - a_i}\left[\frac{x^2}{2}\right]_{a_i}^{b_i} = \frac{b_i^2 - a_i^2}{2(b_i - a_i)} = \frac{a_i + b_i}{2}.\end{aligned} \tag{11b}$$

By words (as it is intuitively obvious): the mean value of the uniform distribution simply is the mean of the lower plus upper limit of the variable range

$$\langle \text{uniform}\_D_i \rangle = \frac{a_i + b_i}{2}. \tag{12a}$$

In order to find the variance of the uniform distribution, we first need finding the second moment

$$\begin{aligned}\langle \text{uniform}\_D_i^2 \rangle &= \int_{a_i}^{b_i} x^2 f_{\text{uniform}\_D_i}(x)dx \\ &= \frac{1}{b_i - a_i}\int_{a_i}^{b_i} x^2\, dx = \frac{1}{b_i - a_i}\left[\frac{x^3}{3}\right]_{a_i}^{b_i} = \frac{b_i^3 - a_i^3}{3(b_i - a_i)} \\ &= \frac{(b_i - a_i)(a_i^2 + a_i b_i + b_i^2)}{3(b_i - a_i)} = \frac{a_i^2 + a_i b_i + b_i^2}{3}.\end{aligned} \tag{12b}$$

The second moment of the uniform distribution is thus

$$\langle \text{uniform}\_D_i^2 \rangle = \frac{a_i^2 + a_i b_i + b_i^2}{3}. \tag{13}$$

From Eqs. (12) and (13), we may now derive the variance of the uniform distribution

$$\begin{aligned}\sigma^2_{\text{uniform}\_D_i} &= \langle \text{uniform}\_D_i^2 \rangle - \langle \text{uniform}\_D_i \rangle^2 \\ &= \frac{a_i^2 + a_i b_i + b_i^2}{3} - \frac{(a_i + b_i)^2}{4} = \frac{(b_i - a_i)^2}{12}.\end{aligned} \tag{14}$$

Upon taking the square root of both sides of Eq. (14), we finally obtain the ***standard deviation of the uniform distribution***:

$$\sigma_{\text{uniform}\_D_i} = \frac{b_i - a_i}{2\sqrt{3}}. \tag{15}$$

We now wish to perform a calculation that is mathematically trivial, but rather unexpected from the intuitive point of view, and very important for our applications to the statistical Drake equation. Just consider the two simultaneous Eqs. (12) and (15)

$$\begin{cases} \langle \text{uniform}\_D_i \rangle = \dfrac{a_i + b_i}{2} \\ \sigma_{\text{uniform}\_D_i} = \dfrac{b_i - a_i}{2\sqrt{3}}. \end{cases} \tag{16}$$

Upon inverting this trivial linear system, one finds

$$\begin{cases} a_i = \langle \text{uniform}\_D_i \rangle - \sqrt{3}\,\sigma_{\text{uniform}\_D_i} \\ b_i = \langle \text{uniform}\_D_i \rangle + \sqrt{3}\sigma_{\text{uniform}\_D_i}. \end{cases} \tag{17}$$

This is of paramount importance for our application the Statistical Drake equation in as much as it shows that:

if one (scientifically) assigns the mean value and standard deviation of a certain Drake random variable Di, then the lower and upper limits of the relevant uniform distribution are given by the two Eq. (17), respectively.

In other words, there is a factor of 3 = 1 : 732 included in the two Eq. (17) that is not obvious at all to human intuition, and must indeed be taken into account.

Proof. of Shannon's 1948 Theorem stating that the Uniform distribution is the "most uncertain" one over any Finite range of values.

As it is well known, the Shannon entropy of any probability density function p(x) is given by the integral

$$\text{Shannon}\_\text{Entropy}\_\text{of}\_p(x) = -\int_{-\infty}^{\infty} p(x) \log p(x)\, dx. \tag{18}$$

In modern textbooks this is also called Shannon differential entropy.

Now consider the case when a probability density function p(x) is limited to a finite interval $a \le x \le b$. This is obviously the case with any physical positive random variable, such as the number N of extraterrestrial communicating civilizations in the Galaxy. We now wish to prove that for any such finite random variable the maximum entropy distribution is the UNIFORM distribution over $a \le x \le b$.

Shannon did not bother to prove this simple theorem in his 1948 papers since he probably regarded it as just too trivial. But we prefer to point out this theorem since, in the language of the statistical Drake equation, it sounds like: "Since we don't know what the probability distribution of any one of the Drake random variables Di is, it is safer to assume that each of them has the maximum possible entropy over $a \le x \le b$, i.e., that Di is UNIFORMLY distributed there".

The proof of this theorem is as follows:

1. Start by assuming $ai \le x \le bi$.
2. Then form the linear combination of the entropy integral plus the normalization condition for Di

$$\delta \int_{a_i}^{b_i} \left[-p(x) \log p(x) + \lambda p(x) \right] dx = 0 \tag{19}$$

where λ is a Lagrange multiplier.

Performing the variation, i.e. differentiating with respect to p(x), one finds

$$-\log p(x) - 1 + \lambda = 0 \tag{20}$$

that is

$$p(x) = e^{\lambda - 1}. \tag{21}$$

Applying the normalization condition (constraint) to the last expression for p(x) yields

$$1 = \int_{a_i}^{b_i} p(x) dx = \int_{a_i}^{b_i} e^{\lambda-1} dx = e^{\lambda-1} \int_{a_i}^{b_i} dx = e^{\lambda-1} (b_i - a_i) \tag{22}$$

that is

$$e^{\lambda-1} = \frac{1}{b_i - a_i} \tag{23}$$

and finally, from (21) and (23)

$$p(x) = \frac{1}{b_i - a_i} \quad \text{with} \quad a_i \leq x \leq b_i. \tag{24}$$

showing that the maximum-entropy probability distribution over any FINITE interval ai ≤ x ≤ bi is just the UNIFORM distribution.

Appendix E

Table 5 Overview of geological ages on planet earth

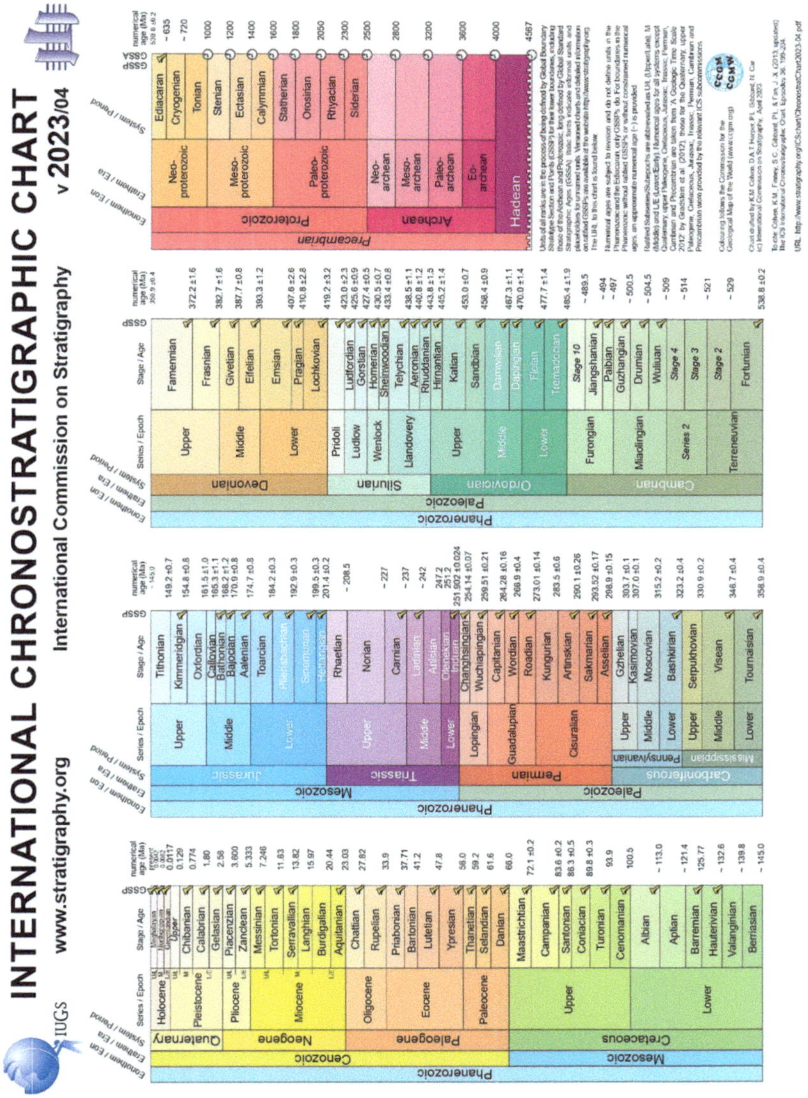

E. Mieli et al., *The Living Galaxy*, https://doi.org/10.1007/978-3-031-67324-5

Bibliography

1. Ageno M (1986) *Le radici della biologia*. Milano, I: Feltrinelli.
2. Ageno M (1991) *Dal non vivente al vivente*. Roma–Napoli, I: Theoria.
3. Altamura E, Albanese P, Marotta R, Milano F, Fiore M, Trotta M, Stano P and Mavelli F (2020) Light–driven ATP production promotes mRNA biosynthesis inside hybrid multi–compartment artificial protocells. *bioRxiv*.
4. Archibald J (2014) *One Plus One Equals One – Symbiosis and the evolution of complex life*. Oxford, UK: Oxford University Press.
5. Arensburg B, Tillier AM, Vandermeersch B, Duday H, Schepartz LA and Rak Y (1989) A Middle Palaeolithic Human Hyoid Bone. *Nature* 338, 758–760.
6. Baker BJ, Tyson GW, Webb RI, Flanagan J, Hugenholtz P, Allen EE and Banfield JF (2006) Lineages of acidophilic archaea revealed by community genomic analysis. *Science* 314, 1933–1935.
7. Baum B and Baum DA (2020) The merger that made us. *BMC Biology* 18, 1–4.
8. Baum DA and Baum B (2014) An Inside–Out origin for the eukaryotic cell. BMC Biology 12, 1–22.
9. Benton MJ (2014) *Vertebrate Palaentology*. New York, NY: John Wiley and Sons, 4th ed.
10. Burcar BT, Barge LM, Trail D, Watson EB, Russell MJ and McGown LB (2015) RNA oligomerization in laboratory analogues of alkaline hydrothermal vent systems, *Astrobiology* 15, 509–522.
11. Butterfield NJ (2004) A vaucheriacean alGy from the middle Neoproterozoic of Spitsbergen: implications for the evolution of Proterozoic eukaryotes and the Cambrian explosion. *Paleobiology* 30, 231–252.
12. Capasso L, Michetti E and D'Anastasio R (2008) A *Homo erectus* hyoid bone: possible implications for the origin of the human capability for speech. *Collegium antropologicum* 32, 1007–1011.

E. Mieli et al., *The Living Galaxy*, https://doi.org/10.1007/978-3-031-67324-5

13. Cavalier-Smith T (2002) Chloroplast evolution: secondary symbiogenesis and multiple losses. Current Biology 12, 62–64.
14. Chavanis P-H, Denet B, Le Berre M, Pomeau Y (2019) Supernova implosion–explosion in the light of catastrophe theory. The European Physical Journal B 92, 1–36.
15. Colín-García M, Heredia A, Cordero G, Camprubí A, Negrón-Mendoza A, Ortega-Gutiérrez F, Beraldi H and Ramos-Bernal S (2016) Hydrothermal vents and prebiotic chemistry: a review. *Boletín de la Sociedad Geológica Mexicana* 68, 599–620.
16. Crossfield IJ, Malik M, Hill ML, Kane SR, Foley B, Alex S, Coria D, Brande J, Zhang Y, Wienke K, Kreidberg L, Cowan NB, Dragomir D, Gorjian V, Mikal-Evans T, Benneke B, Christiansen JL, Deming D and Morales FY (2022) GJ 1252b: A Hot Terrestrial Super–Earth with No Atmosphere. The Astrophysical Journal letters 937, L17.
17. DasGupta S (2020) Molecular crowding and RNA catalysis. *Organic and Biomolecular Chemistry* 18, 7724–7739.
18. de Reviers B (2018) Les associations dans l'évolution du vivant In Palka L (ed), *Microbiodiversité – Un nouveau regard.* Éditions Matériologiques, Paris, pp 51–103.
19. Dodd MS, Papineau D, Grenne T, Slack JF, Rittner M, Pirajno F, O'Neil J and Little CT (2017) Evidence for early life in Earth's oldest hydrothermal vent precipitates. *Nature* 543, 60–64.
20. Doudna JA, Charpentier E (2014) The new frontier of genome engineering with CRISPR–Cas9. Science 346, 1258096.
21. Drake FD (1961) US Academy of Sciences conference on extraterrestrial intelligent life. Green Bank: West Virginia.
22. Egas C, Barroso C, Froufe HJC, Pacheco J, Albuquerque L and da Costa MS (2014) Complete genome sequence of the radiation–resistant bacterium *Rubrobacter radiotolerans* RSPS–4. *Standards in Genomic Sciences* 9, 1062–1075.
23. Eggink LL, Park H and Hoober JK (2001) The role of chlorophyll b in photosynthesis: hypothesis. *BMC Plant Biology* 1, 1–7.
24. El Albani A, Mangano MG, Buatois LA, Bengtson S, Riboulleau A, Bekker A, Konhauser K, Lyons T, Rollion-Bard C, Bankole O, Lekele Baghekema SG, Meunier A, Trentesaux A, Mazurier A, Aubineau J, Laforest C, Fontaine C, Recourt P, Chi Fru E, Macchiarelli E, Reynaud JY, Gauthier-Lafaye F and Canfield DE (2019) Organism motility in an oxygenated shallow–marine environment 2.1 billion years ago. *Proceedings of the National Academy of Sciences* 116, 3431–3436.
25. Eme L and Ettema TJG (2018) The eukaryotic ancestor shapes up. *Nature* 562, 352–353.
26. Erik G (2009) The Milky Way and Beyond Stars. Britannica Educational Publishing.

27. Ettema TJ, Lindås AC and Bernander R (2011) An actin-based cytoskeleton in archaea. *Molecular microbiology* 80, 1052–1061.
28. Fei Y, Seagle CT, Townsend JP, McCoy CA, Boujibar A, Driscoll P, Shulenburger L and Furnish MD (2021) Melting and density of MgSiO3 determined by shock compression of bridgmanite to 1254GPa. Nature Communications 12, 1–9.
29. Ferla MP, Thrash JC, Giovannoni SJ and Patrick WM (2013) New rRNA Gene–Based Phylogenies of the Alphaproteobacteria Provide Perspective on Major Groups, Mitochondrial Ancestry and Phylogenetic Instability. *PLoS ONE* 8, 1–14.
30. Fleagle JG (2013) *Primate Adaptation and Evolution*. London, UK: Academic Press, 3d ed.
31. Flemming HC and Wuertz S (2019) Bacteria and Archaea on Earth and their abundance in biofilms. *Nature Reviews Microbiology* 17, 247–260.
32. Forterre P, Gribaldo S and Brochier C (2005) Luca : à la recherché du plus proche ancêtre commun universel. *Médicine Sciences*, 21, 860–865.
33. Fry I (2000) *The emergency of Life on Earth – a historical and scientific overview*. New Brunswick, NJ: Rutgers University Press.
34. Garwood RJ, Oliver H and Spencer AR (2020) An introduction to the Rhynie chert. *Geological Magazine* 157, 47–64.
35. Gibson TM, Shih PM, Cumming VM, Fischer WW, Crockford PW, Hodgskiss MS, Wörndle S, Creaser RA, Rainbird RH, Skulski TM and Halverson GP (2018) Precise age of *Bangiomorpha pubescens* dates the origin of eukaryotic photosynthesis. *Geology* 46, 135–138.
36. Goëdel K (1931) Über formal unentscheidbare Sätze der Principia Mathematica und verwandter Systeme, I. Monatshefte für Mathematik und Physik 38, 173–198.
37. Gomes R, Levison HF, Tsiganis K and Morbidelli A (2005) Origin of the cataclysmic Late Heavy Bombardment period of the terrestrial planets. Nature 435, 466-469.
38. Gott JR (1993) Implications of the Copernican principle for our future prospects. Nature 363, 315–319.
39. Gould SJ (1989) *Wonderful Life: The Burgess Shale and the Nature of History*. New York, NY: W. W. Norton and Co.
40. Gray MW, Lang BF and Burger G (2004) Mitochondria of protists. *Annual review of genetics* 38, 477–524.
41. Grimaud-Hervé D (1997) *L'évolution de l'encéphale chez* Homo erectus *et* Homo sapiens *: exemples de l'Asie et de l'Europe*. Cahiers de paléoanthropologie. France, F: CNRS Editions.
42. Gros C 2005) Expanding Advanced Civilizations in the Universe. JBIS 58, 1–3
43. Hagadorn JW, Xiao S, Donoghue PC, Bengtson S, Gostling NJ, Pawlowska M, Raff EC, Raff RA, Turner FR, Chongyu Y, Zhou C, Yuan X, McFeely MB, Stampanoni M and Nealson KH (2006) Cellular and subcellular structure of Neoproterozoic animal embryos. *Science* 314, 291–294.

44. Hart MH (1975) An explanation for the absence of extraterrestrials on earth. Quarterly journal of the royal astronomical society 16, 128–135.
45. Holland HD (2006) The oxygenation of the atmosphere and oceans. *Philosophical Transactions of the Royal Society B: Biological Sciences* 361, 903–915.
46. Imachi H, Nobu MK, Nakahara N, Morono Y, Ogawara M, Takaki Y, Takano Y, Uematsu K, Ikuta T, Ito M, Matsui Y, Miyazaki M, Murata K, Saito Y, Sakai S, Song C, Tasumi E, Yamanaka Y, Yamaguchi T, Kamagata Y, Tamaki H and Takai K (2020) Isolation of an archaeon at the prokaryote–eukaryote interface. *Nature* 577, 519–525.
47. Izidoro A (2022) The Exoplanet Radius Valley from Gas–driven Planet Migration and Breaking of Resonant Chains. The Astrophysical Journal Letters 939, L19.
48. Janvier P (1996) Early Vertebrates. Oxford, UK: Clarendon Press.
49. Javaux EJ, Marshall CP and Bekker A (2010) Organic–walled microfossils in 3.2–billionyear–old shallow–marine siliciclastic deposits. *Nature* 463, 934–938.
50. Kardašëv NS (1964) Transmission of Information by Extraterrestrial Civilizations. Soviet Astronomy.
51. Kasting JF (2013) What caused the rise of atmospheric O2? Chemical Geology 362, 13–25.
52. Kauffman SA (2011) Approaches to the origin of life on earth. *Life* 1, 34–48.
53. Kerskens CM and Pérez DL (2022) Experimental indications of non–classical brain functions. Journal of Physics Communications 6, 105001.
54. Knoll AH (2015) *Life in a Young Planet – The first Three Billion years of Evolution on the Earth*. Princeton, NJ: Princeton University Press.
55. Krissansen-Totton J, Arney GN and Catling DC (2018) Constraining the climate and ocean pH of the early Earth with a geological carbon cycle model, *Proceedings of the National Academy of Sciences* 115, 4105–4110.
56. Kunimoto M and Matthews JM (2020) Searching the Entirety of Kepler Data. II. Occurrence Rate Estimates for FGK Stars. The Astronomical Journal 159, 248.
57. Lane N (2002) *Oxygen: the molecule that made the world*. Oxford, UK: Oxford University Press.
58. Lane N (2015) *The Vital Question – Energy, Evolution, and the Origin of the Complex Life*. New York, NY: W. W. Norton and Company.
59. Lane N and Martin W (2010) The energetics of genome complexity. *Nature* 467, p. 929–934.
60. Lecoitre G and Le Guyader H (2017) *Classification phylogénétique du vivant*. Paris, F: Belin, 4th ed.
61. Ledrew G (2001) The Real Starry Sky. Journal of the Royal Astronomical Society of Canada 95, 32.
62. Lei L and Burton ZF (2020) Evolution of life on Earth: tRNA, aminoacyl–tRNA synthetases and the genetic code. *Life* 10, 1–22.
63. Liu Y, Makarova KS, Huang WC, Wolf YI, Nikolskaya AN, Zhang X, Cai M, Zhang C-J, Xu W, Luo Z, Cheng L, Koonin EV and Li M (2021) Expanded

diversity of Asgard archaea and their relationships with eukaryotes. *Nature* 593, 553–557.

64. Lyons TW, Reinhard CT and Planavsky NJ (2014) The rise of oxygen in Earth's early ocean and atmosphere. *Nature* 506, 307–315.
65. Maccone C (2010) The Statistical Drake Equation Acta Astronautica 67, 1366–1383.
66. Maccone C (2015) Statistical Drake–Seager Equation for exoplanet and SETI searches. Acta Astronautica 115, 277–285.
67. Maehara H, Shibayama T, Notsu S, Notsu Y, Nagao T, Kusaba S, Honda S, Nogami D and Shibata K (2012) Superflares on solar–type stars. Nature 485, 478–481.
68. Mallove E and Matloff G (1989) The Starflight Handbook: A Pioneer's Guide to Interstellar Travel. Hoboken, NJ: John Wiley and Sons, Inc.
69. Margalef-Bentabol B, Conselice CJ, Mortlock A, Hartley W, Duncan K, Kennedy R, Kocevski DD, Hasinger G (2018) Stellar populations, stellar masses and the formation of galaxy bulges and discs at z. Monthly Notices of the Royal Astronomical Society 473, 5370–5384.
70. Marguet E, Gaudin M, Gauliard E, Fourquaux I, Plouy S, Matsui I and Forterre P (2013) Membrane vesicles, nanopods and/or nanotubes produced by hyperthermophilic archaea of the genus *Thermococcus*. *Biochemical Society Transactions* 41, 436–442.
71. Margulis L (1998) *Symbiotic planet – A new look at evolution*. New York, NY: Basic Books.
72. Martin W and Müller M (1998) The hydrogen hypothesis for the first eukaryote. *Nature* 392, 37–41.
73. McHenry HM and Coffing K (2000) *Australopithecus* to *Homo*: transformations in body and mind. *Annual review of Anthropology* 29, 125–146.
74. McMenamin MAS (1998) *The garden of Ediacara – discovering the first complex life*. New York, NY: Columbia University Press.
75. McShea DW (2001) The hierarchical structure of organisms: a scale and documentation of a trend in the maximum. *Paleobiology* 27, 405–423.
76. Mendell JE, Clements KD, Choat JH and Angert ER (2008) Extreme polyploidy in a large bacterium. *Proceedings of the National Academy of Sciences* 105, 6730–6734.
77. Ménez B, Pisapia C, Andreani M, Jamme F, Vanbelligen QP, Brunelle A, Richard L, Dumas P and Réfrégiers M (2018) Abiotic synthesis of amino acids in the recesses of the oceanic lithosphere. *Nature* 564, 59–63.
78. Mikhailov KV, Konstantinova AV, Nikitin MY, Troshin PV, Rusin LY, Lyubetsky VA, Panchin YV, Mylnikov AP, Moroz LL, Kumar S and Aleoshin VV (2009) The origin of Metazoa: a transition from temporal to spatial cell differentiation. *Bioessays* 31, 758–768.
79. Miller S (1953) A production of amino acids under possible primitive Earth conditions. *Science* 117, 528–529.

80. Moyà-Solà S, Köhler M and Rook L (2005) The *Oreopithecus* thumb: a strange case in hominoid evolution. *Journal of human evolution* 49, 395–404.
81. Neveu M, Kim HJ and Benner SA (2013) The "strong" RNA world hypothesis: Fifty years old. *Astrobiology* 13, 391–403.
82. Och LM and Shields-Zhou GY (2012) The Neoproterozoic oxygenation event: Environmental perturbations and biogeochemical cycling. *Earth–Science Reviews* 110, 26–57.
83. Ogunseitan OA (2016) Bacterial Diversity, Introduction to In Kliman RM (ed) *Encyclopedia of Evolutionary Biology (vol. 1)*. Oxford, UK: Academic Press, pp 114–118.
84. Papineau D, She Z, Dodd MS, Iacoviello F, Slack JF, Hauri E, Shearing P and Little CTS (2022) Metabolically diverse primordial microbial communities in Earth's oldest seafloor–hydrothermal jasper. *Science Advances* 8, 1–16.
85. Parfrey LW, Lahr DJG, Knoll AH and Katz LA (2011) Estimating the timing of early eukaryotic diversification with multigene molecular clocks. *Proceedings of the National Academy of Sciences* 108, 13624–13629.
86. Penrose R (1990) The Emperor's New Mind: Concerning Computers, Minds, and the Laws of Physics. Oxford, UK: Oxford University Press.
87. Penrose R (1994) Shadows of the Mind: A Search for the Missing Science of Consciousness. Oxford, UK: Oxford University Press.
88. Porter SM (2004) The fossil record of early eukaryotic diversification. *The Paleontological Society Papers* 10, 35–50.
89. Rasmussen B, Fletcher IR, Brocks JJ and Kilburn MR (2008) Reassessing the first appearance of eukaryotes and cyanobacteria. *Nature* 455, 1101–1104.
90. Raup DM (1992) *Extinction – Bad Genes or Bad Luck?* New York, NY: WW Norton and Company.
91. Ritson D and Sutherland JD (2012) Prebiotic synthesis of simple sugars by photoredox systems chemistry. *Nature Chemistry* 4, 895–899.
92. Rospars J-P (2013) Trends in the evolution of life, brains and intelligence. *International Journal of Astrobiology* 12, 186–207.
93. Russell DA and Séguin R (1982) Reconstruction of the small Cretaceous theropod *Stenonychosaurus inequalis* and a hypothetical dinosauroid. *Syllogeus* 37,1–43.
94. Sagan C and Drake F (1975) The Search for Extraterrestrial Intelligence. *Scientific American* 232, 80–89.
95. Schirrmeister BE, Sanchez-Baracaldo P, and Wacey D (2016) Cyanobacterial evolution during the Precambrian. *Cyanobacterial evolution during the Precambrian* 15, 187–204.
96. Schoch RR (2014) *Amphibian Evolution – The Life of Early Land Vertebrate.* Oxford, UK: WILEY Blackwell.
97. Schopf JM and Parcker BM (1987) Early Archean (3.3 billion to 3.5 billionyear old) microfossil from Warrawoona Group, Australia, *Science* 237, 70–73.
98. Schopf JW, Kudryavtsev AB, Osterhout JT, Williford KH, Kitajima K, Valley JW and Sugitani K. (2017) An anaerobic~ 3400 My shallow-water microbial

consortium: Presumptive evidence of Earth's Paleoarchean anoxic atmosphere, *Precambrian Research* 299, p. 309–318.
99. Sebé-Pedrós A, Degnan BM and Ruiz-Trillo I (2017) The origin of Metazoa: a unicellular perspective. *Nature Reviews Genetics* 18, 498–512.
100. Shevchenko II., Melnikov AV, Popova EA, Bobylev VV and Karelin GM (2019) Circumbinary Planetary Systems in the Solar Neighborhood: Stability and Habitability. Astronomy Letters 45, 620–626.
101. Smulsky JJ (2011) The Influence of the Planets, Sun and Moon on the Evolution of the Earth's Axis, International Journal of Astronomy and Astrophysics 1, 117.
102. Sojo V, Herschy B, Whicher A, Camprubí E and Lane N (2016) The Origin of Life in Alkaline Hydrothermal Vents. *Astrobiology* 16, p. 181–200.
103. Southam G, Rothschild LJ and Westall F (2007) The geology and habitability of terrestrial planets: fundamental requirements for life. *Space Science Reviews* 129, 7–34.
104. Spang A, Saw JH, Jørgensen SL, Zaremba-Niedzwiedzka K, Martijn J, Lind AE, van Eijk R, Schleper C, Guy L and Ettema TJG (2015) Complex archaea that bridge the gap between prokaryotes and eukaryotes. *Nature* 521, 173–179.
105. Stanford CB (2001) *The Hunting Apes: Meat Eating and the Origins of Human Behavior*. Princeton, NJ: Princeton University Press.
106. Stworzewicz E, Szulc J and Pokryszko BM (2009) Late Paleozoic continental gastropods from Poland: Systematic, evolutionary and paleoecological approach. *Journal of Paleontology* 83, 938–945.
107. Summons RE, Bradley AS, Jahnke LL and Waldbauer JR (2006) Steroids, triterpenoids and molecular oxygen. *Philosophical Transactions of the Royal Society B: Biological Sciences* 361, 951–968.
108. Susman RL (2005) *Oreopithecus*: still apelike after all these years. *Journal of human evolution* 49, 405–411.
109. Sweeney D, Tuthill P, Sharma S and Hirai R (2022) The Galactic underworld: the spatial distribution of compact remnants. Monthly Notices of the Royal Astronomical Society 516, 4971–4979.
110. Tattersall I (2016) A tentative framework for the acquisition of language and modern human cognition. *Journal of Anthropological Sciences* 94, 157–166.
111. Taylor TN, Taylor EL and Krings M (2009) *Paleobotany: The Biology and Evolution of Fossil Plants*. Amsterdam, HO: Academic Press, 2d ed.
112. van Tuinen M and Hadly EA (2004) Error in Estimation of Rate and Time Inferred from the Early Amniote Fossil Record and Avian Molecular Clocks. *Journal of Molecular Evolution* 59, 267–276.
113. Vinge VS (1993) Technological Singularity. In VISION-21 Symposium sponsored by NASA Lewis Research Center and the Ohio Aerospace Institute. Vernor Vinge Magazine: Whole Earth Review, pp. 30-31.
114. Webb S (2015) *If the Universe is Teeming with Aliens... where is Everybody? – Seventy–Five Solutions to the Fermi Paradox and the Problem of Extraterrestrial Life*. Berlin, GE: Springer International Publishing, 2nd ed.

115. Yamaguchi M, Mori Y, Kozuka Y, Okada H, Uematsu K, Tame A, Furukawa H, Maruyama T, O'Driscoll Worman C and Yokoyama K (2012) Prokaryote or eukaryote? A unique microorganism from the deep sea. *Journal of Electron Microscopy* 61, 423–431.
116. Zhang B (2018) The Physics of Gamma–Ray Bursts. Cambridge, UK: Cambridge University Press.
117. Zhao W, Zhang X, Jia G, Shen YA and Zhu M (2021) The Silurian–Devonian boundary in East Yunnan (South China) and the minimum constraint for the lungfish–tetrapod split. *Science China Earth Sciences* 64, 1784–1797.
118. Zimorski V, Mentel M, Tielens AG and Martin WF (2019) Energy metabolism in anaerobic eukaryotes and Earth's late oxygenation, Free Radical Biology and Medicine, 140, p. 279–294.
119. Cohen KM, Finney SC, Gibbard PL and Fan J-X (2013; updated) The ICS International Chronostratigraphic Chart. *Episodes* 36, 199–204.
120. Takeuchi Y, Furukawa Y, Kobayashi T, Sekine T, Terada N and Kakegawa T (2020) Impact–induced amino acid formation on Hadean Earth and Noachian Mars. *Scientific Reports* 10, 1–7.

MIX
Papier aus verantwortungsvollen Quellen
Paper from responsible sources
FSC® C105338

If you have any concerns about our products,
you can contact us on
ProductSafety@springernature.com

In case Publisher is established outside the EU,
the EU authorized representative is:
Springer Nature Customer Service Center GmbH
Europaplatz 3, 69115 Heidelberg, Germany

Printed by Libri Plureos GmbH
in Hamburg, Germany